PILOT TRAINING MANUAL FOR THE B-17 FLYING FORTRESS

USAAF

1944

S.O.A. BOOKS

Published by S.O.A. Books, Canberra

First promulgated by U.S.A.A.F. in 1944

ISBN: **978-1-925907-11-7**

This edition © S.O.A. Books, 2024, all rights reserved. No part of this publication may be reproduced, stored in a retrieval system or transmitted, in any form or by any means, electronic, mechanical, photocopying, recording or otherwise without prior permission from the publisher.

RESTRICTED

PILOT TRAINING MANUAL FOR THE
Flying Fortress
B-17

PUBLISHED FOR HEADQUARTERS, AAF

OFFICE OF ASSISTANT CHIEF OF AIR STAFF, TRAINING

BY HEADQUARTERS, AAF, OFFICE OF FLYING SAFETY

Foreword

THIS MANUAL is the text for your training as a B-17 pilot and airplane commander.

The Air Forces' most experienced training and supervisory personnel have collaborated to make it a complete exposition of what your pilot duties are, how each duty will be performed, and why it must be performed in the manner prescribed.

The techniques and procedures described in this book are standard and mandatory. In this respect the manual serves the dual purpose of a training checklist and a working handbook. Use it to make sure that you learn everything described herein. Use it to study and review the essential facts concerning everything taught. Such additional self-study and review will not only advance your training, but will alleviate the burden of your already overburdened instructors.

This training manual does not replace the Technical Orders for the airplane, which will always be your primary source of information concerning the B-17 so long as you fly it. This is essentially the textbook of the B-17. Used properly, it will enable you to utilize the pertinent Technical Orders to even greater advantage.

GENERAL U. S. ARMY,
COMMANDING GENERAL,
ARMY AIR FORCES

THE FIRST FORTRESS: The Air Corps called for a "battleship of the skies;" Boeing offered the "299" (later the XB-17); observers called it a "regular fortress with wings." It exceeded expectations, later crashed—victim of pilot error.

THE STORY OF THE B-17

In 1934 the U. S. Army Air Corps asked for a battleship of the skies. The specifications called for a "multi-engine" bomber that would have a high speed of 200-250 mph at 10,000 feet, an operating speed of 170-200 mph at the same altitude, a range of 6 to 10 hours, and a service ceiling of 20,000-25,000 feet.

Boeing designers figured that with a conventional 2-engine type of airplane they could meet all specifications and probably better them. But such a design probably would not provide much edge over the entries of competitors. They decided to build a revolutionary type of 4-engine bomber.

In July 1935 an airplane such as the world had never seen before rolled out on the apron of the Boeing plant at Seattle, Wash. It was huge: 105 feet in wing span, 70 feet from nose to tail, 15 feet in height, It was equipped with 4 Pratt & Whitney Hornet 750 Hp engines, and 4 Hamilton Standard 3-bladed constant-speed propellers. To eliminate air resistance, its bomb load was tucked away in internal bomb bays. Pilots and crew had soundproofed, heated, comfortable quarters where they could operate efficiently while flying in any kind of climate. And the big bomber bristled with formidable firepower.

"It's a regular fortress," someone observed, "a fortress with wings."

Thus the Boeing 299, later designated the XB-17, was born—the grandfather of the Flying Fortress that was to become champion and pace-setter of all heavy bombardment aircraft in the World War II.

The XB-17 surpassed all Army specifications

NEXT CAME THE Y1B-17: Thirteen were delivered in 1937. One stalled, spun down over Langley Field, recovered, landed safely. Recording instruments showed it had held up under greater stress than it was designed to stand.

THE B-17A WENT HIGHER: Equipped with turbos, it topped all previous service ceilings, gave maximum performance above 30,000 ft. To range, speed, bomb load, firepower, the B-17 added another advantage: altitude operation.

for speed, climb, range, and load-carrying requirements. Then, in October, 1935, it crashed at Wright Field when a test pilot neglected to unlock the elevator controls on takeoff.

But the Army Air Corps recognized in this first Fortress the heavy bomber of the future. Thirteen airplanes, designated Y1B-17, were ordered. While one airplane was held at Wright Field for experimental purposes, the other 12 went out to set new range and speed records, cruising the Western Hemisphere, and confounding skeptics who said that the Flying Fortress was "too much airplane for any but super-pilots." Not one of the 12 was ever destroyed by accident.

With experience, the Fortress acquired new strength, virtues, possibilities. The Y1B-17A, equipped with Wright G Cyclone engines and General Electric turbo-superchargers, gave astonishing performances at altitudes above 30,000 feet. The B-17B, flight tested in 1939, had 1000 Hp Wright Cyclone engines and hydromatic full-feathering propellers.

In the spring of 1940, when Hitler had overrun Norway, Denmark, Holland, Belgium, and France, the B-17C made its debut with more armor plate for crew protection, more power in its engines. The B-17D took on leakproof fuel tanks, increased armament, better engine cooling in fast climbs, and a speed increase to more than 300 mph.

When the Japs attacked Pearl Harbor, the B-17C's and B-17D's were the first Fortresses to see action. But soon the B-17E's were on their way to join them in even greater numbers—faster, heavier, sturdier Fortresses, packing .50-cal. waist and tail guns, with a Sperry ball turret under the fuselage, and another power turret on top.

By the spring of 1942, still another Fortress—the B-17F—with longer range, greater bomb load capacity, more protective armament and striking power, was streaking across both Atlantic and Pacific in enormous numbers to provide what General Arnold called "the guts and backbone of our world-wide aerial offensive."

THE FIRST B-17B left Seattle 1 August, 1939, arrived in New York 9 hours, 14 minutes later, setting a new coast-to-coast non-stop record. Later, seven B-17B's cruised the hemisphere for the 50th anniversary of the Republic of Brazil.

RESTRICTED

THE B-17D SAW ACTION FIRST: When the Japs struck, Fortresses of the C and D series gained experience that later made the B-17 the "guts and backbone of our worldwide aerial offensive." B-17E was first wartime model.

THE B-17F FULFILLED THE PROMISE: With over 400 major changes—producing greater speed, range, bomb load, firepower, crew protection—new Forts swept the Pacific and the heart of Europe, raised the curtain on D-Day.

Rugged Forts Make History

The combat record of the Flying Fortress has been written daily in newspaper headlines since Dec. 7, 1941.

From the hour of Pearl Harbor, through the dark, early months of the war in the Pacific, they were sinking Jap ships and shooting arrogant Zeros out of the skies.

They carried the war to the enemy in the Coral Sea, over Guadalcanal, New Guinea, Java, Burma, and the Bismarck Sea.

Changing tactics, they hedgehopped volcanic peaks, flew practically at water level through unbroken fog, to bomb the Japs out of the Aleutians.

They flew the blistering deserts to drive the enemy out of North Africa, the Mediterranean, Sicily, and open the way to Rome.

Pilot points proudly to battle-scarred Fortress—calls it "series of holes held together by ragged metal."

Doomed ME 109 plowed into this Flying Fortress over Tunisia, cutting fuselage nearly in half, entirely removing one elevator. Pilot nursed the airplane home to British base, brought it in for a perfect landing.

RESTRICTED

Beginning in August, 1942, they brought daylight bombing to Hitler's Europe, first over strategic targets in Occupied France, then gradually spreading out over the continent until, in the spring of 1944, shuttle bombing from bases in Britain and Russia left no corner of the once haughty Festung Europa safe from concentrated Allied bombing attacks.

Detailed Fortress history must remain a voluminous post-war job for military historians. For pilots, however, one important fact stands clear-cut now. The Flying Fortress is a rugged airplane.

In the words of one veteran: "She'll not only get you to the target and do the job, but she'll fight her way out, take terrific punishment, and get you safely home."

Headlines have reiterated that fact with heart-warming redundancy:

40 NAZIS RIDDLE FORT, BUT FAIL TO DOWN IT.

LAME FORTRESS BAGS 6 GERMANS, MAKES HOME BASE.

B-17, SPLIT IN TWO, LANDS SAFELY.

FORT FALLS 10,000 FEET, BUT COMPLETES RAID.

FORT LIMPS HOME ON ONE MOTOR.

HARD-HIT FORT CUTS LOOSE BALL TURRET, GETS HOME.

The ground crew looked up and saw, coming down for a landing, not the Flying Fortress, but a lone motor. Sitting on the motor was a sergeant with a machine gun across his lap. He brought the motor down to a beautiful no-point landing, jumped off.

"Boy," he said, "were we in a fight!"
—Yank: The Army Weekly

With only ragged pieces of tail left, this Fortress, believed wrecked in enemy territory, limped home.

RESTRICTED

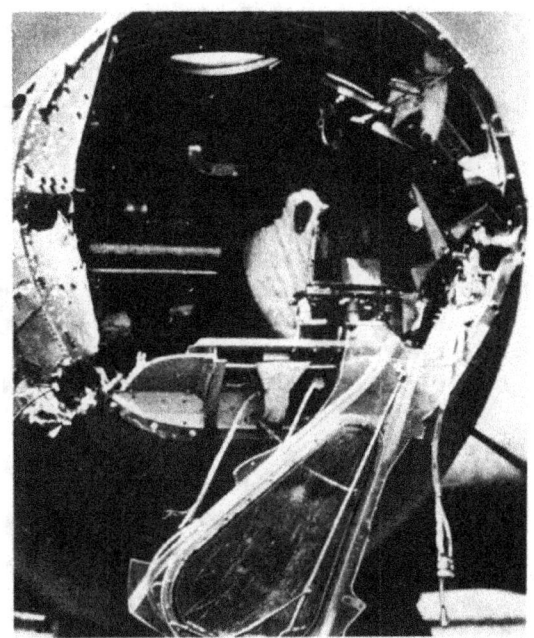

Swarms of FW 190's shot out plexiglas nose, killed navigator, wounded entire crew. Pilot brought in airplane safely despite loss of flaps, hydraulic system.

Gaping flak and shell holes, received while bombing Nazi aircraft factories, failed to prevent "F for Frenesi" from returning safely with three wounded gunners.

Fragments of German 8.3-inch rocket tore into fuselage aft of cockpit, made long journey home difficult. Most Fortress crews, always amazed by airplane's ability to take it, have a word for the fighting B-17: "Rugged."

RESTRICTED

FORT STRUGGLES HOME WITH TAIL BLOWN OFF.

The B-17's incredible capacity to take it—to come flying home on three, two, even one engine, sieve-like with flak and bullet holes, with large sections of wing or tail surfaces shot away—has been so widely publicized that U. S. fighting men could afford to joke about it.

But the fact remains: the rugged Forts can take it and still fly home. Why?

The B-17 is built for battle. Its wings are constructed with heavy truss-type spars which tend to localize damage by enemy fire so that basic wing strength is not affected.

Because of its unusual tail design, the airplane can be flown successfully even when vertical or horizontal tail surfaces have been partially destroyed, or with one or more engines shot away.

Even when battle damage prevents use of all other control methods, the autopilot provides near-normal maneuverability.

There are many other reasons. But perhaps the most important of all is the fact that every man who flies one knows that the B-17 is a pilot's airplane. It inspires confidence and warrants it. For the fulfillment of its intended function it demands just one thing: pilot know-how.

DUTIES AND RESPONSIBILITIES OF
THE AIRPLANE COMMANDER

Your assignment to the B-17 airplane means that you are no longer just a pilot. You are now an airplane commander, charged with all the duties and responsibilities of a command post.

You are now flying a 10-man weapon. It is your airplane, and your crew. You are responsible for the safety and efficiency of the crew at all times—not just when you are flying and fighting, but for the full 24 hours of every day while you are in command.

Your crew is made up of specialists. Each man—whether he is the navigator, bombardier, engineer, radio operator, or one of the gunners—is an expert in his line. But how well he does his job, and how efficiently he plays his part as a member of your combat team, will depend to a great extent on how well you play your own part as the airplane commander.

Get to know each member of your crew as an individual. Know his personal idiosyncrasies, his capabilities, his shortcomings. Take a personal interest in his problems, his ambitions, his need for specific training.

See that your men are properly quartered, clothed, and fed. There will be many times, when your airplane and crew are away from the home base, when you may even have to carry your interest to the extent of financing them yourself. Remember always that you are the commanding officer of a miniature army—a specialized army; and that morale is one of the biggest problems for the commander of any army, large or small.

Crew Discipline

Your success as the airplane commander will depend in a large measure on the respect, confidence, and trust which the crew feels for you. It will depend also on how well you maintain crew discipline.

Your position commands obedience and respect. This does not mean that you have to be stiff-necked, overbearing, or aloof. Such characteristics most certainly will defeat your purpose.

Be friendly, understanding, but firm. Know your job; and, by the way you perform your duties daily, impress upon the crew that you do know your job. Keep close to your men, and let them realize that their interests are uppermost in your mind. Make fair decisions, after due consideration of all the facts involved; but make them in such a way as to impress upon your crew that your decisions are to stick.

Crew discipline is vitally important, but it need not be as difficult a problem as it sounds. Good discipline in an air crew breeds comradeship and high morale, and the combination is unbeatable.

You can be a good CO, and still be a regular guy. You can command respect from your men, and still be one of them.

"To associate discipline with informality, comradeship, a leveling of rank, and at times a shift in actual command away from the leader, may seem paradoxical," says a brigadier general, formerly a Group commander in the VIII Bomber Command. "Certainly, it isn't down the military groove. But it is discipline just the same—and the kind of discipline that brings success in the air."

Crew Training

Train your crew as a team. Keep abreast of their training. It won't be possible for you to follow each man's courses of instruction, but you can keep a close check on his record and progress.

Get to know each man's duties and problems. Know his job, and try to devise ways and means of helping him to perform it more efficiently.

Each crew member naturally feels great pride in the importance of his particular specialty. You can help him to develop his pride to include the manner in which he performs that duty. To do that you must possess and maintain a thorough knowledge of each man's job and the problems he has to deal with in the performance of his duties.

THE COPILOT

The copilot is the executive officer—your chief assistant, understudy, and strong right arm. He must be familiar enough with every one of your duties—both as pilot and as airplane commander—to be able to take over and act in your place at any time.

He must be able to fly the airplane under all conditions as well as you would fly it yourself.

He must be extremely proficient in engine operation, and know instinctively what to do to keep the airplane flying smoothly even though he is not handling the controls.

He must have a thorough knowledge of cruising control data, and know how to apply it at the proper time.

He is also the engineering officer aboard the airplane, and maintains a complete log of performance data.

He must be a qualified instrument pilot.

He must be able to fly good formation in any assigned position, day or night.

He must be qualified to navigate by day or at night by pilotage, dead reckoning, and by use of radio aids.

He must be proficient in the operation of all radio equipment located in the pilot's compartment.

In formation flying, he must be able to make engine adjustments almost automatically.

He must be prepared to take over on instruments when the formation is climbing through an overcast, thus enabling you to watch the rest of the formation.

Always remember that the copilot is a fully trained, rated pilot just like yourself. He is subordinate to you only by virtue of your position as the airplane commander. The B-17 is a lot of airplane; more airplane than any one pilot can handle alone over a long period of time. Therefore, you have been provided with a second pilot who will share the duties of flight operation.

Treat your copilot as a brother pilot. Remember that the more proficient he is **as a pilot**, the more efficiently he will be able to perform the duties of the vital post he holds as your second in command.

Be sure that he is allowed to do his share of the flying, in the pilot's seat, on takeoffs, landings, and on instruments.

The importance of the copilot is eloquently testified by airplane commanders overseas. There have been many cases in which the pilot has been disabled or killed in flight and the copilot has taken full command of both airplane and crew, completed the mission, and returned safely to the home base. Usually, the copilots who have distinguished themselves under such conditions have been copilots who have been respected and trained by the airplane commander **as pilots**.

Bear in mind that the pilot in the right-hand seat of your airplane is preparing himself for an airplane commander's post too. Allow him every chance to develop his ability and to profit by your experience.

THE NAVIGATOR

The navigator's job is to direct your flight from departure to destination and return. He must know the exact position of the airplane at all times.

Navigation is the art of determining geographic positions by means of (a) pilotage, (b) dead reckoning, (c) radio, or (d) celestial navigation, or any combination of these 4 methods. By any one or combination of methods the navigator determines the position of the airplane in relation to the earth.

Pilotage

Pilotage is the method of determining the airplane's position by visual reference to the ground. The importance of accurate pilotage cannot be over-emphasized. In combat navigation, all bombing targets are approached by pilotage, and in many theaters the route is maintained by pilotage. This requires not merely the vicinity type, but **pin-point pilotage**. The exact position of the airplane must be known not within 5 miles but within ¼ of a mile.

The navigator does this by constant reference to groundspeeds and ETA's established for points ahead, the ground, and to his maps and charts. During the mission, so long as he can maintain visual contact with the ground, the navigator can establish these pin-point positions so that the exact track of the airplane will be known when the mission is completed.

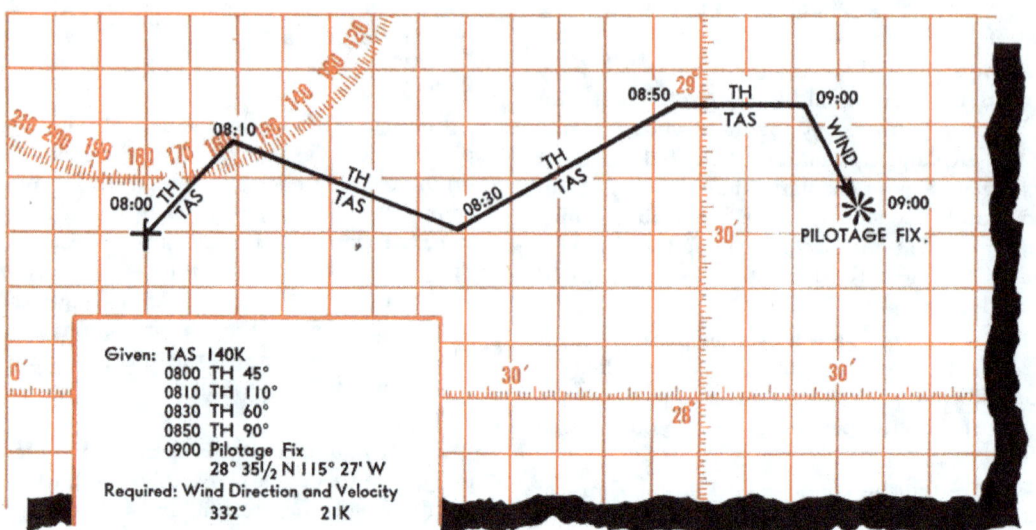

Dead Reckoning

Dead reckoning is the basis of all other types of navigation. For instance, if the navigator is doing pilotage and computes ETA's for points ahead, he is using dead reckoning.

Dead reckoning determines the position of the airplane at any given time by keeping an account of the track and distance flown over the earth's surface from the point of departure or the last known position.

Dead reckoning can be subdivided into two classes:

1. **Dead reckoning as a result of a series of known positions obtained by some other means of navigation.** For example, you, as pilot, start on a mission from London to Berlin at 25,000 feet. For the first hour your navigator keeps track by pilotage; at the same time recording the heading and airspeed which you are holding. According to plan, at the end of the first hour the airplane goes above the clouds, thus losing contact with the ground. By means of dead reckoning from his last pilotage point, the navigator is able to tell the position of the aircraft at any time. The first hour's travel has given him the wind prevalent at altitude, and the track and groundspeed being made. By computing track and distance from the last pilotage point, he can always tell the position of the airplane. When your airplane comes out of the clouds near Berlin, the navigator will have a very close approximation of his exact position, and will be able to pick up pilotage points quickly.

2. **Dead reckoning as a result of visual references other than pilotage.** When flying over water, desert, or barren land, where no reliable pilotage points are available, accurate DR navigation still can be performed. By means of the drift meter the navigator is able to determine drift, the angle between the heading of the airplane and its track over the ground. The true heading of the airplane is obtained by application of compass error to the compass reading. The true heading plus or minus the drift (as read on the drift meter) gives the track of the airplane. At a constant airspeed, drift on 2 or more headings will give the navigator information necessary to obtain the wind by use of his computer. Groundspeed is computed easily once the wind, heading, and airspeed are known. So, by constant recording of true heading, true airspeed, drift, and groundspeed, the navigator is able to determine accurately the position of the airplane at any given time. For greatest accuracy, the pilot must maintain constant courses and airspeeds. If course or airspeed is changed, notify the navigator so he can record these changes.

Radio

Radio navigation makes use of various radio aids to determine position. The development of many new radio devices has increased the use of radio in combat zones. However, the ease with which radio aids can be jammed, or bent, limits the use of radio to that of a check on DR and pilotage. The navigator, in conjunction with the radio man, is responsible for all radio procedures, approaches, etc., that are in effect in the theater.

Celestial

Celestial navigation is the science of determining position by reference to 2 or more celestial bodies. The navigator uses a sextant, accurate time, and many tables to obtain what he calls a line of position. Actually this line is part of a circle on which the altitude of the particular body is constant for that instant of time. An intersection of 2 or more of these lines gives the navigator a fix. These fixes can be relied on as being accurate within approximately 10 miles. One reason for inaccuracy is the instability of the airplane as it moves through space, causing acceleration of the sextant bubble (a level denoting the horizontal). Because of this acceleration, the navigator takes observations over a period of time so that the acceleration error will cancel out to some extent. If the navigator tells the pilot when he wishes to take an observation, extremely careful flying on the part of the pilot during the few minutes it takes to make the observation will result in much greater accuracy. Generally speaking, the only celestial navigation used by a combat crew is during the delivering flight to the theater. But in all cases celestial navigation is used as a check on dead reckoning and pilotage except where celestial is the only method available, such as on long over-water flights, etc.

Instrument Calibration

Instrument calibration is an important duty of the navigator. All navigation depends directly on the accuracy of his instruments. Correct calibration requires close cooperation and extremely careful flying by the pilot. Instruments to be calibrated include the altimeter, all compasses, airspeed indicators, alignment of the astrocompass, astrograph, and drift meter, and check on the navigator's sextant and watch.

Pilot-Navigator Preflight Planning

1. Pilot and navigator must study flight plan of the route to be flown and select alternate airfields.
2. Study the weather with the navigator. Know what weather you are likely to encounter. Decide what action is to be taken. Know the weather conditions at the alternate airfields.
3. Inform your navigator at what airspeed and altitude you wish to fly so that he can prepare his flight plan.
4. Learn what type of navigation the navigator intends to use: pilotage, dead reckoning, radio, celestial, or a combination of all methods.
5. Determine check points; plan to make radio fixes.
6. Work out an effective communication method with your navigator to be used in flight.
7. Synchronize your watch with your navigator's.

Pilot-Navigator in Flight

1. **Constant course**—For accurate navigation, the pilot—you—must fly a constant course. The navigator has many computations and entries to make in his log. Constantly changing course makes his job more difficult. A good navigator is supposed to be able to follow the pilot, but he cannot be taking compass readings all the time.
2. **Constant airspeed** must be held as nearly as possible. This is as important to the navigator as is a constant course in determining position.
3. **Precision flying** by the pilot greatly affects the accuracy of the navigator's instrument readings, particularly celestial readings. A slight error in celestial reading can cause considerable error in determining positions. You can help the navigator by providing as steady a platform as possible from which he can take readings. The navigator should notify you when he intends to take readings so that the airplane can be leveled off and flown as smoothly as possible, preferably by using the automatic pilot.

Do not allow your navigator to be disturbed while he is taking celestial readings.

4. **Notify the navigator of any change in flight,** such as change in altitude, course, or airspeed. If change in flight plan is to be made, consult the navigator. Talk over the proposed change so that he can plan the flight and advise you about it.

5. If there is doubt about the position of the airplane, pilot and navigator should get together, refer to the navigator's flight log, talk the problem over and decide together the best course of action to take.

6. Check your compasses at intervals with those of the navigator, noting any deviation.

7. Require your navigator to give position reports at intervals.

8. You are ultimately responsible for getting the airplane to its destination. Therefore, it is your duty to know your position at all times.

9. Encourage your navigator to use as many navigation methods as possible as a means of double-checking.

Post-flight Critique

After every flight, get together with the navigator and discuss the flight and compare notes. Go over the navigator's log. If there have been serious navigational errors, discuss them with the navigator and determine their cause. If the navigator has been at fault, caution him that it is his job to see that the same mistake does not occur again. If the error has been caused by faulty instruments, see that they are corrected before another navigation mission is attempted. If your flying has contributed to inaccuracy in navigation, try to fly a better course next time.

Miscellaneous Duties

The navigator's primary duty is navigating your airplane with a high degree of accuracy. But as a member of the team, he must also have a general knowledge of the entire operation of the airplane.

He has a .50-cal. machine gun at his station, and he must be able to use it skillfully and to service it in emergencies.

He must be familiar with the oxygen system, know how to operate the turrets, radio equipment, and fuel transfer system.

He must know the location of all fuses and spare fuses, lights and spare lights, affecting navigation.

He must be familiar with emergency procedures, such as the manual operation of landing gear, bomb bay doors, and flaps, and the proper procedures for crash landings, ditching, bailout, etc.

THE BOMBARDIER

Accurate and effective bombing is the ultimate purpose of your entire airplane and crew. Every other function is preparatory to hitting and destroying the target.

That's your bombardier's job. The success or failure of the mission depends upon what he accomplishes in that short interval of the bombing run.

When the bombardier takes over the airplane for the run on the target, he is in absolute command. He will tell you what he wants done, and until he tells you "Bombs away," his word is law.

A great deal, therefore, depends on the understanding between bombardier and pilot. You expect your bombardier to know his job when he takes over. He expects you to understand the problems involved in his job, and to give him full cooperation. Teamwork between pilot and bombardier is essential.

Under any given set of conditions—groundspeed, altitude, direction, etc.—there is only one point in space where a bomb may be released from the airplane to hit a predetermined object on the ground.

There are many things with which a bombardier must be thoroughly familiar in order to release his bombs at the right point to hit this predetermined target.

He must know and understand his bombsight, what it does, and how it does it.

He must thoroughly understand the operation and upkeep of his bombing instruments and equipment.

He must know that his racks, switches, controls, releases, doors, linkage, etc., are in first-class operating condition.

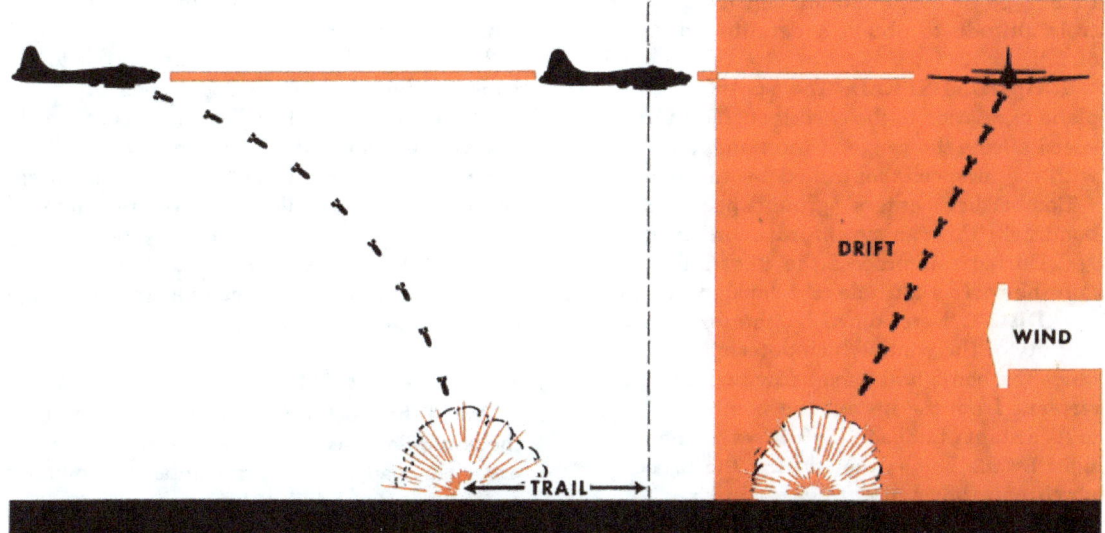

He must understand the automatic pilot as it pertains to bombing.

He must know how to set it up, make any adjustments and minor repairs while in flight.

He must know how to operate all gun positions in the airplane.

He must know how to load and clear simple stoppages and jams of machine guns while in flight.

He must be able to load and fuse his own bombs.

He must understand the destructive power of bombs and must know the vulnerable spots on various types of targets.

He must understand the bombing problem, bombing probabilities, bombing errors, etc.

He must be thoroughly versed in target identification and in aircraft identification.

The bombardier should be familiar with the duties of all members of the crew and should be able to assist the navigator in case the navigator becomes incapacitated.

For the bombardier to be able to do his job, the pilot of the aircraft must place the aircraft in the proper position to arrive at a point on a circle about the target from which the bombs can be released to hit the target.

Consider the following conditions which affect the bomb dropped from an airplane:—

1. **ALTITUDE:** Controlled by the pilot. Determines the length of time the bomb is sustained in flight and affected by atmospheric conditions, thus affecting the range (forward travel of the bomb) and deflection (distance the bomb drifts in a crosswind with respect to airplane's ground track).

2. **TRUE AIRSPEED:** Controlled by the pilot. The measure of the speed of the airplane through the air. It is this speed which is imparted to the bomb and which gives the bomb its initial forward velocity and, therefore, affects the trail of the bomb, or the distance the bomb lags behind the airplane at the instant of impact.

3. **BOMB BALLISTICS:** Size, shape and density of the bomb, which determines its air resistance. Bombardier uses bomb ballistics tables to account for type of bomb.

4. **TRAIL:** Horizontal distance the bomb is behind the airplane at the instant of impact. This value, obtained from bombing tables, is set in the sight by the bombardier. Trail is affected by altitude, airspeed, bomb ballistics and air density, the first 2 factors being controlled by the pilot.

5. **ACTUAL TIME OF FALL:** Length of time the bomb is sustained in air from instant of release to instant of impact. Affected by alti-

tude, type of bomb and air density. Pilot controls altitude to obtain a definite actual time of fall.

6. **GROUNDSPEED:** The speed of the airplane in relation to the earth's surface. Groundspeed affects the range of the bomb and varies with the airspeed, controlled by the pilot.

Bombardier enters groundspeed in the bombsight through synchronization on the target. During this process the pilot must maintain the correct altitude and constant airspeed.

7. **DRIFT:** Determined by the direction and velocity of the wind, which determines the distance the bomb will travel downwind from the airplane from the instant the bomb is released to its instant of impact. Drift is set on the bombsight by the bombardier during the process of synchronization and setting up course.

The above conditions indicate that the pilot plays an important part in determining the proper point of release of the bomb. Moreover, throughout the course of the run, as explained below, there are certain preliminaries and techniques which the pilot must understand to insure accuracy and minimum loss of time.

Prior to takeoff the pilot must ascertain that the airplane's flight instruments have been checked and found accurate. These are the altimeter, airspeed indicator, free air temperature gauge and all gyro instruments. These instruments must be used to determine accurately the airplane's attitude.

The Pilot's Preliminaries

The autopilot and PDI should be checked for proper operation. It is very important that PDI and autopilot function perfectly in the air; otherwise it will be impossible for the bombardier to set up an accurate course on the bombing run. The pilot should thoroughly familiarize himself with the function of both the C-1 autopilot and PDI.

If the run is to be made on the autopilot, the pilot must carefully adjust the autopilot before reaching the target area. The autopilot must be adjusted under the same conditions that will exist on the bombing run over the target. For this reason the following factors should be taken into consideration and duplicated for initial adjustment.

1. Speed, altitude and power settings at which run is to be made.

2. Airplane trimmed at this speed to fly hands off with bomb bay doors opened.

The same condition will exist during the actual run, except that changes in load will occur before reaching the target area because of gas consumption. The pilot will continue making adjustments to correct for this by disengaging the autopilot elevator control and re-trimming the airplane, then re-engaging and adjusting the autopilot trim of the elevator.

Setting Up the Autopilot

One of the most important items in setting up the autopilot (see pp. 185-188) for bomb approach is to adjust the turn compensation knobs so that a turn made by the bombardier will be coordinated and at constant altitude. Failure to make this adjustment will involve difficulty and delay for the bombardier in establishing an accurate course during the run—with the possibility that the bombardier may not be able to establish a proper course in time, the result being considerably large deflection errors in point of impact.

Uncoordinated turns by the autopilot on the run cause erratic lateral motion of the course hair of the bombsight when sighting on target. The bombardier in setting up course must eliminate any lateral motion of the fore-and-aft hair in relation to the target before he has the proper course set up. Therefore, any erratic motion of the course hair requires an additional correction by the bombardier, which would not be necessary if autopilot was adjusted to make coordinated turns.

USE OF THE PDI: The same is true if PDI is used on the bomb run. Again, coordinated smooth turns by the pilot become an essential part of the bomb run. In addition to added course corrections necessitated by uncoordinated turns, skidding and slipping introduce small changes in airspeed affecting synchronization of the bombsight on the target. To help the pilot flying the run on PDI, the airplane should be trimmed to fly practically hands off.

Assume that you are approaching the target area with autopilot properly adjusted. Before reaching the initial point (beginning of bomb run) there is evasive action to be considered. Many different types of evasive tactics are employed, but from experience it has been recommended that the method of evasive action be left up to the bombardier, since the entire antiaircraft pattern is fully visible to the bombardier in the nose.

EVASIVE ACTION: Changes in altitude necessary for evasive action can be coordinated with the bombardier's changes in direction at specific intervals. This procedure is helpful to the bombardier since he must select the initial point at which he will direct the airplane onto the briefed heading for the beginning of the bomb run.

Should the pilot be flying the evasive action on PDI (at the direction of the bombardier) he must know the exact position of the initial point for beginning the run, so that he can fly the airplane to that point and be on the briefed heading. Otherwise, there is a possibility of beginning to run too soon, which increases the airplane's vulnerability, or beginning the run too late, which will affect the accuracy of the bombing. For best results the approach should be planned so the airplane arrives at the initial point on the briefed heading, and at the assigned bombing altitude and airspeed.

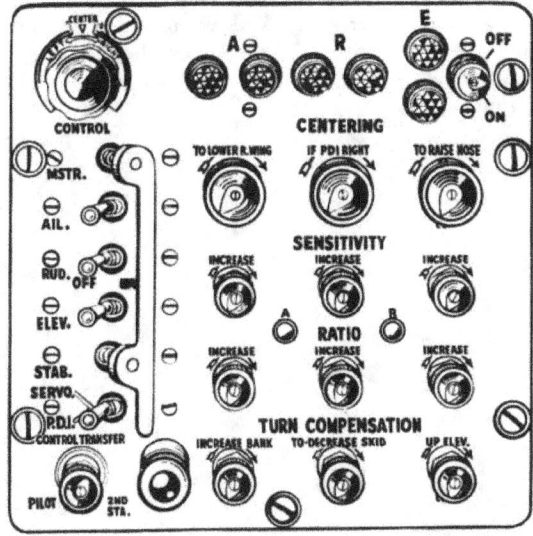

AUTOPILOT

PDI

At this point the bombardier and pilot as a team should exert an extra effort to solve the problem at hand. It is now the bombardier's responsibility to take over the direction of flight, and give directions to the pilot for the operations to follow. The pilot must be able to follow the bombardier's directions with accuracy and minimum loss of time, since the longest possible bomb run seldom exceeds 3 minutes. Wavering and indecision at this moment are disastrous to the success of any mission, and during the crucial portion of the run, flak and fighter opposition must be ignored if bombs are to hit the target. The pilot and bombardier should keep each other informed of anything which may affect the successful completion of the run.

HOLDING A LEVEL: Either before or during the run, the bombardier will ask the pilot for a level. This means that the pilot must accurately level his airplane with his instruments (ignoring the PDI). There should be no acceleration of the airplane in any direction, such as an increase or decrease in airspeed, skidding or slipping, gaining or losing altitude.

For the level the pilot should keep a close check on his instruments, not by feel or watching the horizon. Any acceleration of the airplane during this moment will affect the bubbles (through centrifugal force) on the bombsight gyro, and the bombardier will not be able to establish an accurate level.

For example, assume that an acceleration occurred during the moment the bombardier was accomplishing a level on the gyro. A small increase in airspeed or a small skid, hardly perceptible, is sufficient to shift the gyro bubble liquid 1° or more. An erroneous tilt of 1° on the gyro will cause an error of approximately 440 feet in the point of impact of a bomb dropped from 20,000 feet, the direction of error depending on direction of tilt of gyro caused by the erroneous bubble reading.

HOLDING ALTITUDE AND AIRSPEED: As the bombardier proceeds to set up his course (synchronize), it is absolutely essential that the pilot maintain the selected altitude and airspeed within the closest possible limits. For every additional 100 feet above the assumed 20,000-foot bombing altitude, the bombing error will increase approximately 30 feet, the direction of error being over. For erroneous airspeed, which creates difficulty in synchronization on the target, the bombing error will be approximately 170 feet for a 10 mph change in airspeed. Assuming the airspeed was 10 mph in excess, from 20,000 feet, the bomb impact would be short 170 feet.

The pilot's responsibility to provide a level and to maintain a selected altitude and airspeed within the closest limits cannot be over-emphasized.

If the pilot is using PDI (at the direction of the bombardier) instead of autopilot, he must be thoroughly familiar with the corrections demanded by the bombardier. Too large a correction or too small a correction, too soon or too late, is as bad as no correction at all. Only through prodigious practice flying with the PDI can the pilot become proficient to a point where he can actually perform a coordinated turn, the amount and speed necessary to balance the bombardier's signal from the bombsight.

Erratic airspeeds, varying altitudes, and poorly coordinated turns make the job of establishing course and synchronizing doubly difficult for both pilot and bombardier, because of the necessary added corrections required. The resulting bomb impact will be far from satisfactory.

After releasing the bombs, the pilot or bombardier may continue evasive action—usually the pilot, so that the bombardier may man his guns.

The pilot using the turn control may continue to fly the airplane on autopilot, or fly it manually, with the autopilot in a position to be engaged by merely flipping the lock switches. This would provide potential control of the airplane in case of emergency.

REDUCING CIRCULAR ERROR: One of the greatest assets towards reducing the circular error of a bombing squadron lies in the pilot's ability to adjust the autopilot properly, fly the PDI, and maintain the designated altitude and airspeeds during the bombing run. Reducing the circular error of a bombing squadron reduces the total number of aircraft required to destroy a particular target. For this reason both pilot and bombardier should work together until they have developed a complete understanding and confidence in each other.

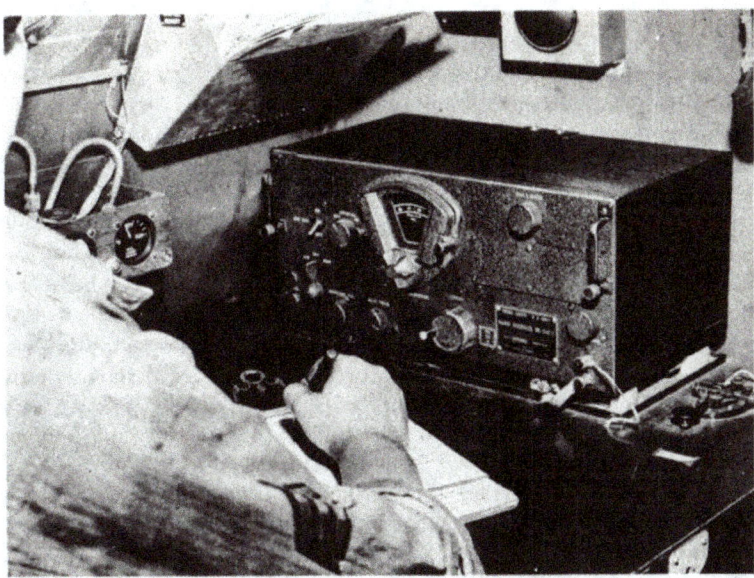

THE RADIO OPERATOR

There is a lot of radio equipment in today's B-17's. There is one man in particular who is supposed to know all there is to know about this equipment. Sometimes he does, but often he doesn't. And when the radio operator's deficiencies do not become apparent until the crew is in the combat zone, it is then too late. Too often the lives of pilots and crew are lost because the radio operator has accepted his responsibility indifferently.

Radio is a subject that cannot be learned in a day. It cannot be mastered in 6 weeks, but sufficient knowledge can be imparted to the radio man during his period of training in the United States if he is willing to study. It is imperative that you check your radio operator's ability to handle his job before taking him overseas as part of your crew. To do this you may have to check the various departments to find any weakness in the radio operator's training and proficiency and to aid the instructors in overcoming such weaknesses.

Training in the various phases of the heavy bomber program is designed to fit each member of the crew for the handling of his jobs. The radio operator will be required to:

1. Render position reports every 30 minutes.
2. Assist the navigator in taking fixes.
3. Keep the liaison and command sets properly tuned and in good operating order.
4. Understand from an operational point of view:
   (a) Instrument landing
   (b) IFF
   (c) VHF

and other navigational aids equipment in the airplane.

5. Maintain a log.

In addition to being a radio operator, the radio man is also a gunner. During periods of combat he will be required to leave his watch at the radio and take up his guns. He is often required to learn photography. Some of the best pictures taken in the Southwest Pacific were taken by radio operators. The radio operator who cannot perform his job properly may be the weakest member of your crew—and the crew is no stronger than its weakest member.

THE ENGINEER

Size up the man who is to be your engineer. This man is supposed to know more about the airplane you are to fly than any other member of the crew.

He has been trained in the Air Forces' highly specialized technical schools. Probably he has served some time as a crew chief. Nevertheless, there may be some inevitable blank spots in his training which you, as a pilot and airplane commander, may be able to fill in.

Think back on your own training. In many courses of instruction, you had a lot of things thrown at you from right and left. You had to concentrate on how to fly; and where your equipment was concerned you learned to rely more and more on the enlisted personnel, particularly the crew chief and the engineer, to advise you about things that were not taught to you because of lack of time and the arrangement of the training program.

Both pilot and engineer have a responsibility to work closely together to supplement and fill in the blank spots in each other's education.

To be a qualified combat engineer a man must know his airplane, his engines, and his armament equipment thoroughly. This is a big responsibility: the lives of the entire crew, the safety of the equipment, the success of the mission depend upon it squarely.

He must work closely with the copilot, checking engine operation, fuel consumption, and the operation of all equipment.

He must be able to work with the bombardier, and know how to cock, lock, and load the bomb racks. It is up to you, the airplane commander, to see that he is familiar with these duties, and, if he is hazy concerning them, to have the bombardier give him special help and instruction.

He must be thoroughly familiar with the armament equipment, and know how to strip, clean, and re-assemble the guns.

He should have a general knowledge of radio equipment, and be able to assist in tuning transmitters and receivers.

Your engineer should be your chief source of information concerning the airplane. He should know more about the equipment than any other crew member—yourself included.

You, in turn, are his source of information concerning flying. Bear this in mind in all your discussions with the engineer. The more complete you can make his knowledge of the reasons behind every function of the equipment, the more valuable he will be as a member of the crew. Who knows? Someday that little bit of extra knowledge in the engineer's mind may save the day in some emergency.

Generally, in emergencies, the engineer will be the man to whom you turn first. Build up his pride, his confidence, his knowledge. Know him personally; check on the extent of his knowledge. Make him a man upon whom you can rely.

THE GUNNERS

The B-17 is a most effective gun platform, but its effectiveness can be either applied or defeated by the way the gunners in your crew perform their duties in action.

Your gunners belong to one of two distinct categories: turret gunners and flexible gunners.

The power turret gunners require many mental and physical qualities similar to what we know as inherent flying ability, since the operation of the power turret and gunsight are much like that of airplane flight operation.

While the flexible gunners do not require the same delicate touch as the turret gunner, they must have a fine sense of timing and be familiar with the rudiments of exterior ballistics.

All gunners should be familiar with the coverage area of all gun positions, and be prepared to bring the proper gun to bear as the conditions may warrant.

They should be experts in aircraft identification. Where the Sperry turret is used, failure to set the target dimension dial properly on the K-type sight will result in miscalculation of range.

They must be thoroughly familiar with the Browning aircraft machine gun. They should know how to maintain the guns, how to clear jams and stoppages, and how to harmonize the sights with the guns.

While participating in training flights, the gunners should be operating their turrets constantly, tracking with the flexible guns even when actual firing is not practical. Other airplanes flying in the vicinity offer excellent tracking targets, as do automobiles, houses, and other ground objects during low altitude flights.

The importance of teamwork cannot be overemphasized. One poorly trained gunner, or one man not on the alert, can be the weak link as a result of which the entire crew may be lost.

Keep the interest of your gunners alive at all times. Any form of competition among the gunners themselves should stimulate interest to a high degree.

Finally, each gunner should fire the guns at each station to familiarize himself with the other man's position and to insure knowledge of operation in the event of an emergency.

General Description

THIS IS THE FLYING FORTRESS, B-17F: A 4-ENGINE, MID-WING MONOPLANE OF ALL-METAL, ALUMINUM ALLOY, STRESSED-SKIN CONSTRUCTION.

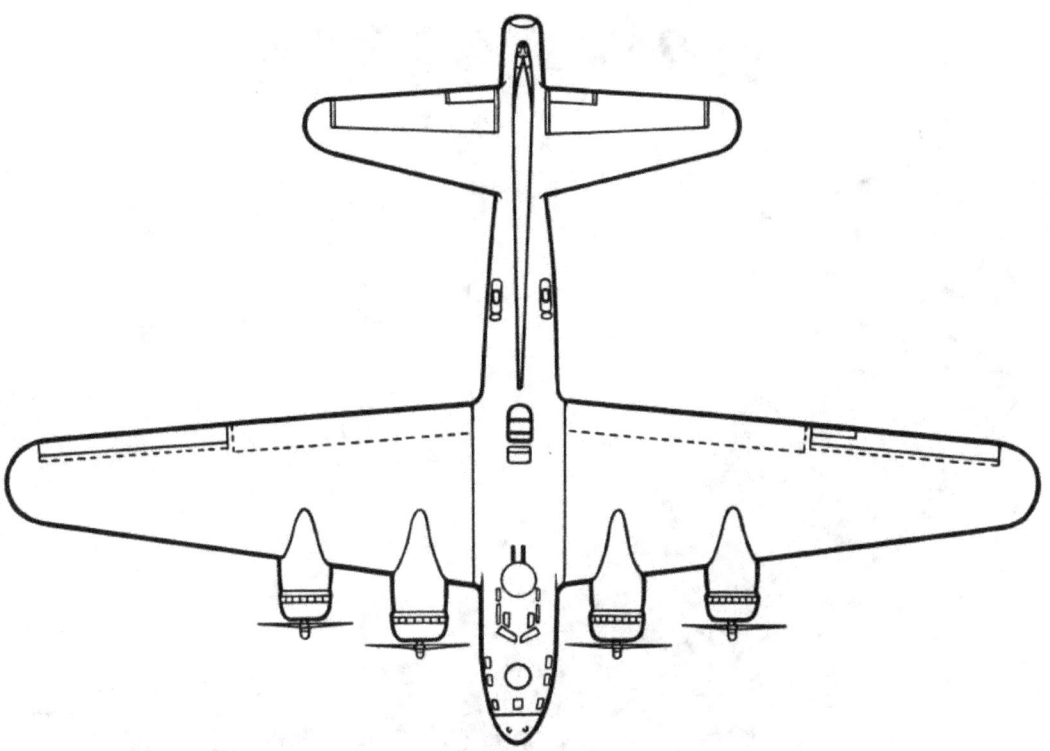

APPROXIMATE OVER-ALL DIMENSIONS:
Length: 74 feet, 9 inches
Height: 19 feet, 1 inch (gear down)
Wing Span: 103 feet, 9 inches

APPROXIMATE WEIGHT:
Tactical empty: 41,000 lb.
Maximum gross: 64,500 lb.

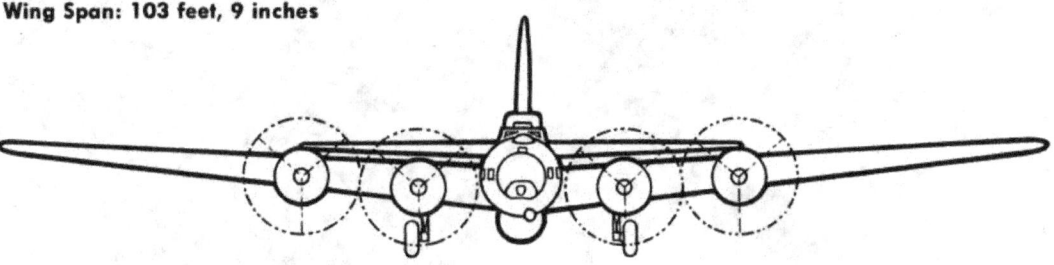

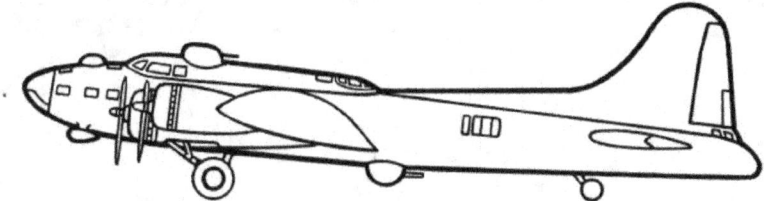

Construction

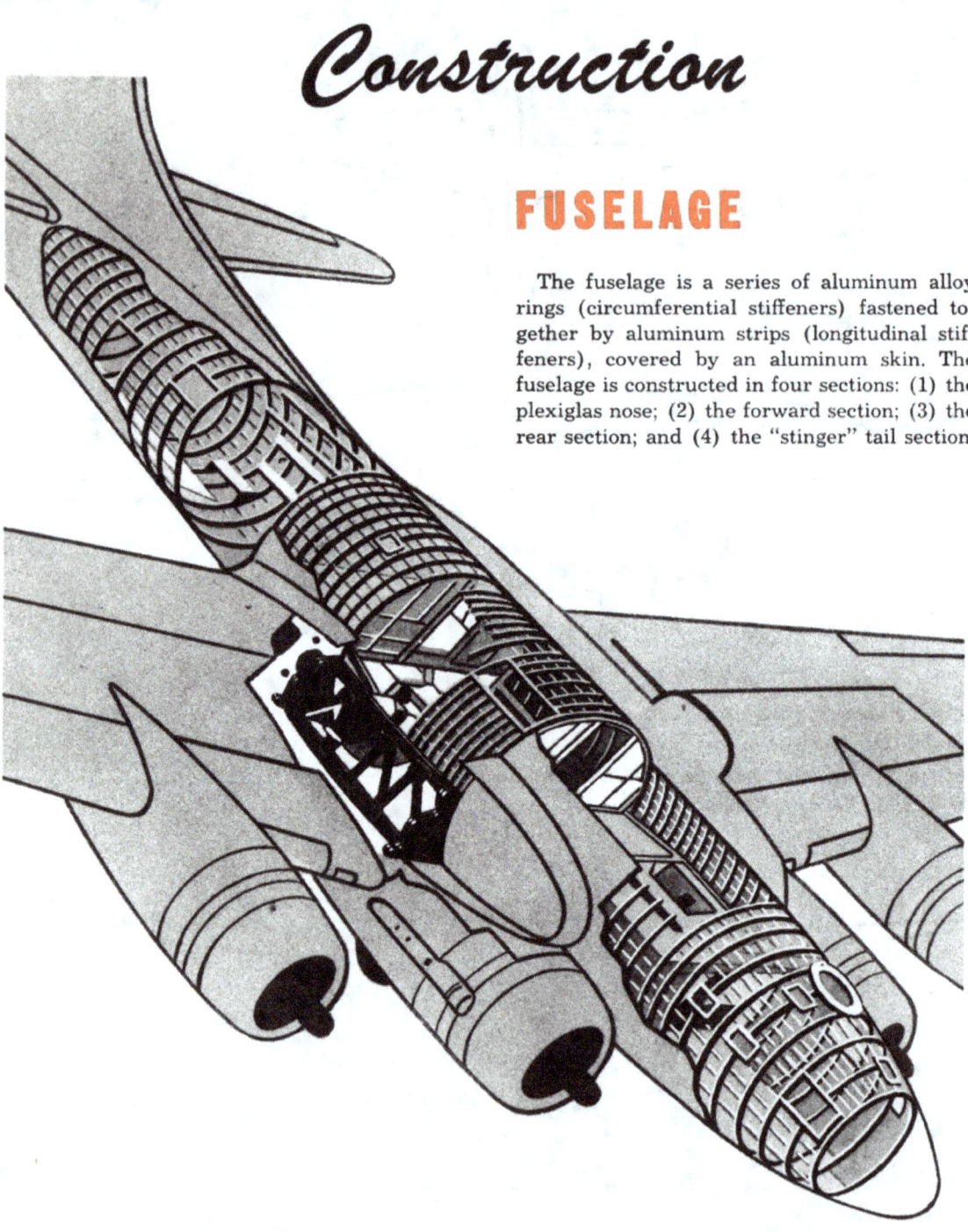

FUSELAGE

The fuselage is a series of aluminum alloy rings (circumferential stiffeners) fastened together by aluminum strips (longitudinal stiffeners), covered by an aluminum skin. The fuselage is constructed in four sections: (1) the plexiglas nose; (2) the forward section; (3) the rear section; and (4) the "stinger" tail section.

TAIL ASSEMBLY

RESTRICTED

Tail surfaces, both vertical and horizontal, are similar in structure to the wings, except that sheet stiffeners, instead of corrugated sheet, are used to support the skin. The loads on these surfaces are lighter, hence the structure is made comparatively lighter.

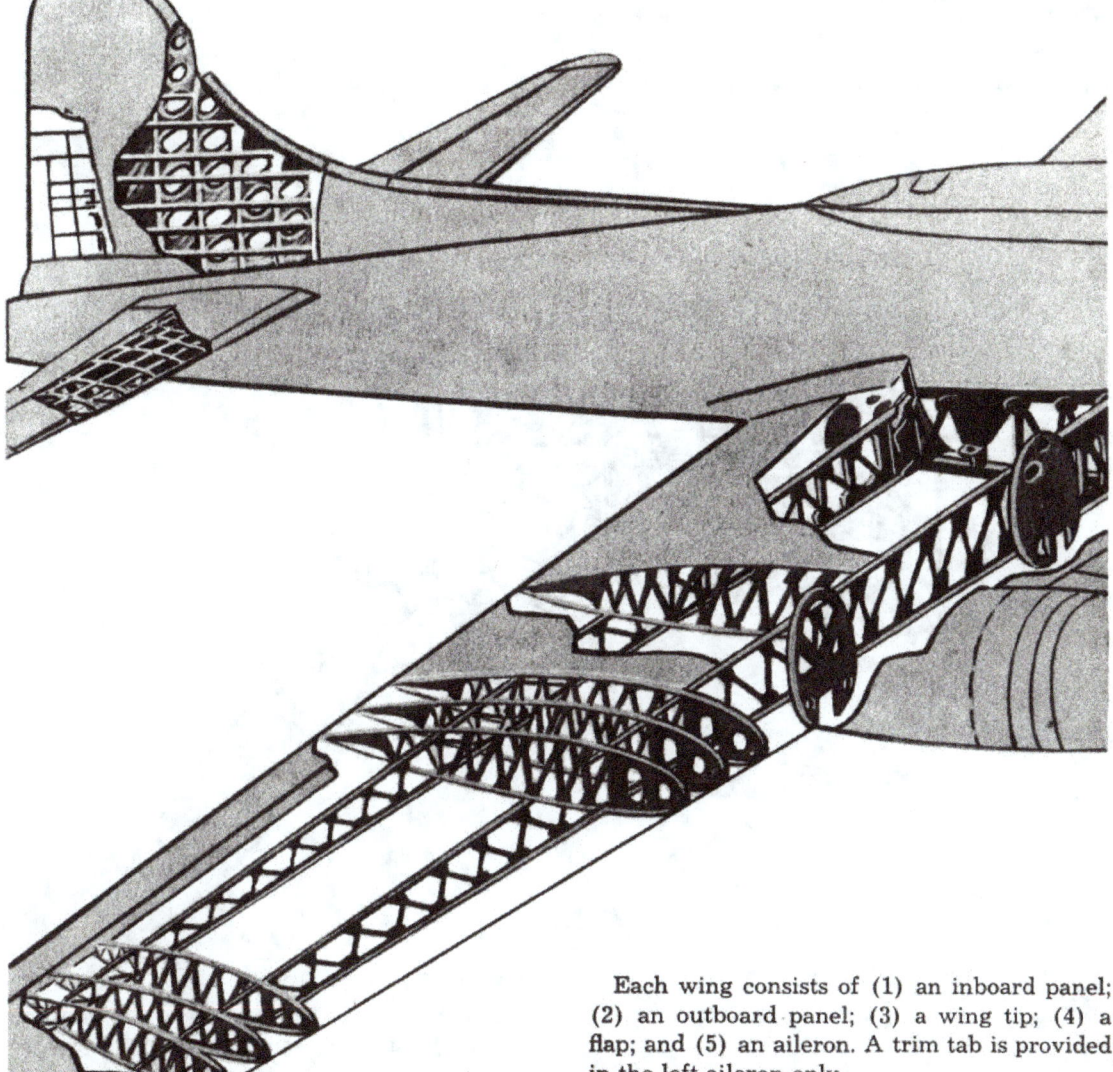

WINGS

Each wing consists of (1) an inboard panel; (2) an outboard panel; (3) a wing tip; (4) a flap; and (5) an aileron. A trim tab is provided in the left aileron only.

The engine nacelles, of semi-monocoque design, are installed in each inboard wing panel.

The wing construction: spars and highly stressed ribs of the truss type. Corrugated dural sheet, attached to the rib cords, reinforces the skin to withstand compression loads.

POWER PLANT

Engines

The B-17F has four 1200 Hp Wright Cyclone Model R-1820-97 engines of the 9-cylinder, radial, air-cooled type with a 16-to-9 gear ratio.

Each engine has a turbo-supercharger to boost manifold pressure for takeoff and maintain sea-level pressure at high altitude.

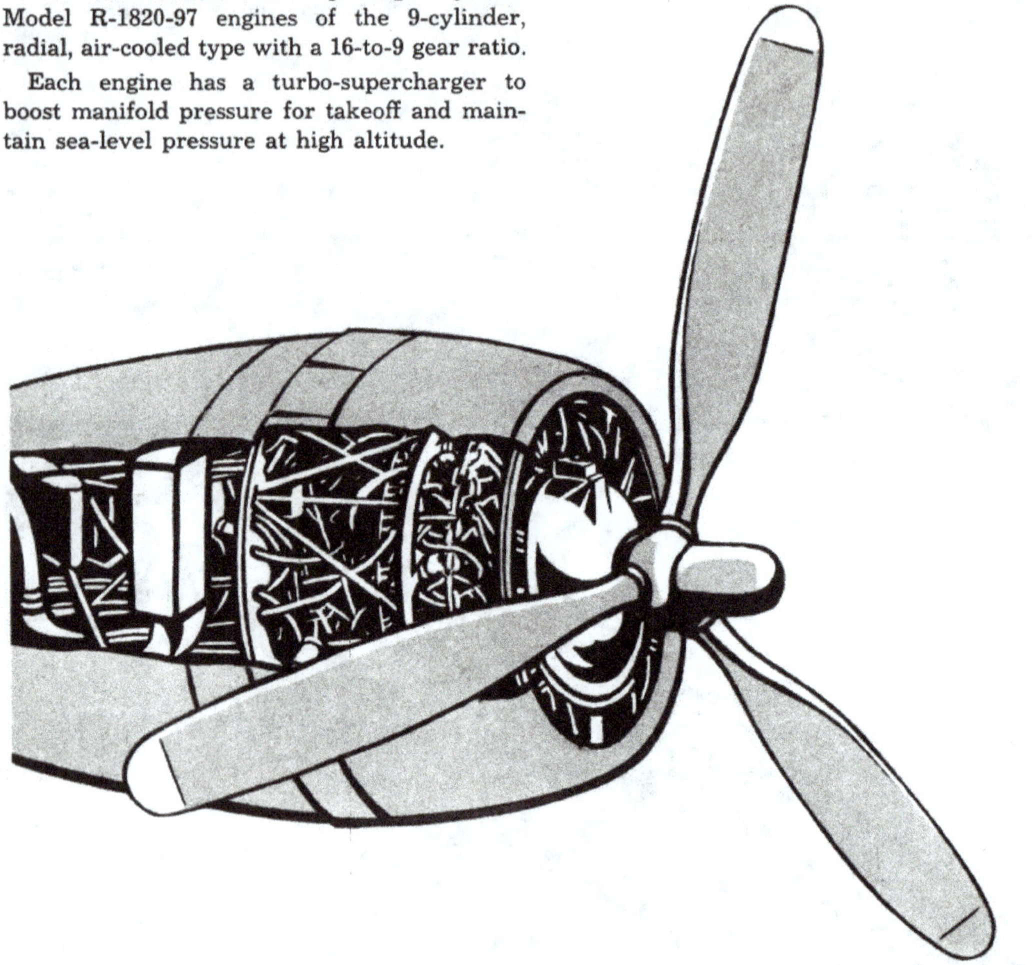

Propellers

The Hamilton Standard 3-bladed propellers are hydromatically controlled with constant-speed and full feathering provisions. Adjustment of the propeller governors is accomplished individually by cable controls from the cockpit. Feathering and unfeathering is accomplished hydraulically by an electric motor-driven pump mounted on the forward side of the firewall in each engine nacelle.

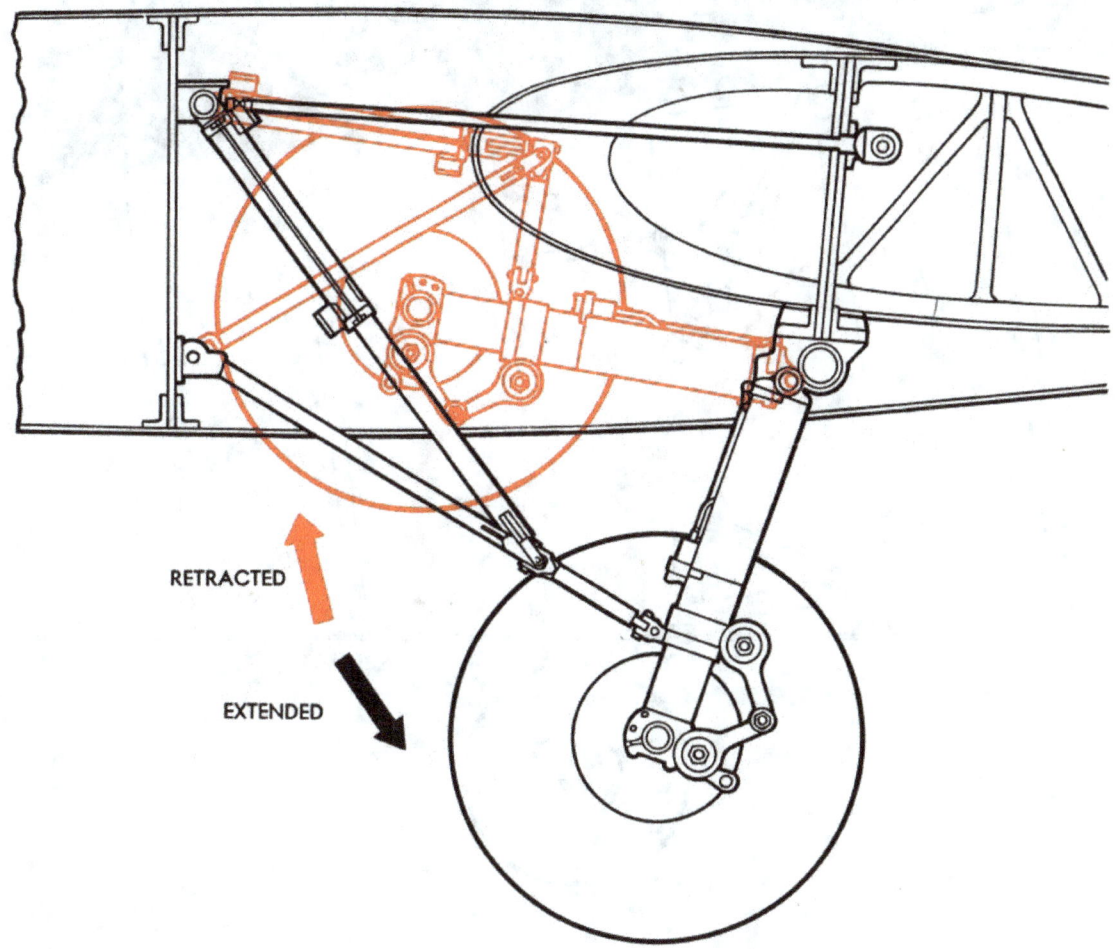

RETRACTED

EXTENDED

MAIN LANDING GEAR

The B-17F landing gear is of the conventional type: left hand and right hand main gear and a tail gear.

The main gear retracts into the nacelles behind the inboard engines. Electrically operated retraction units, with auxiliary manual systems, are used for raising and lowering the main wheels. The emergency hand crank connections for operating the main landing gear are at the forward end of the bomb bay on each side of the doorway leading to the cockpit.

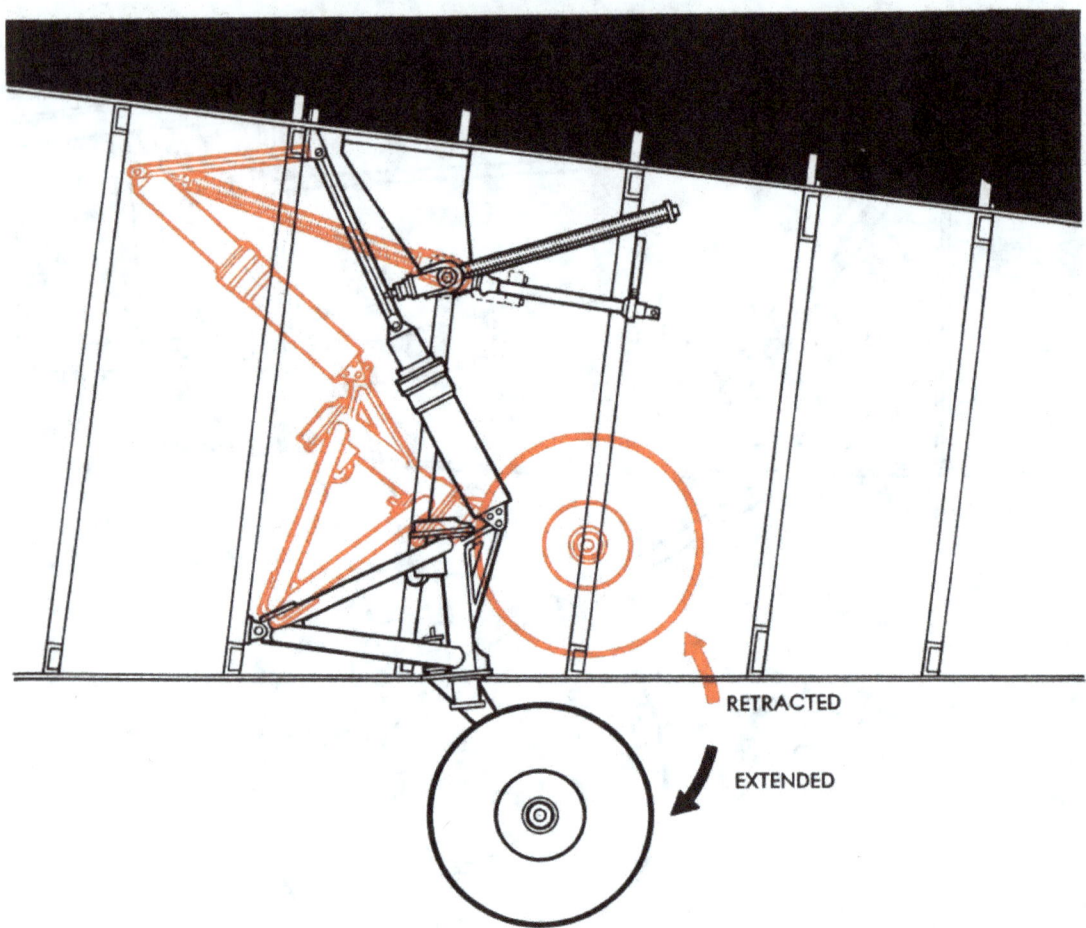

RETRACTED

EXTENDED

TAILWHEEL

The tail gear consists of a wheel assembly, knuckle, treadle, oleo, retraction unit, anti-shimmy brake and wheel lock. Provisions are made for 360° rotation of the wheel, and for locking the wheel in a straight fore-and-aft position during takeoff. The tailwheel gear may be retracted either electrically or manually. Electrical retraction is controlled in the cockpit with the same toggle switch that controls the main landing gear retraction motor. For manual retraction, a hand crank is operated through the motor slip clutch.

Interior of the Airplane

PILOT'S COMPARTMENT

Between the nose section and the bomb bay is the flight deck, or pilot's compartment. This elevated enclosure contains the pilot's and co-pilot's stations with all the essential flight controls, instruments, etc., (See pp. 34-39.) It is also equipped with a Sperry power turret with twin .50-cal. machine guns.

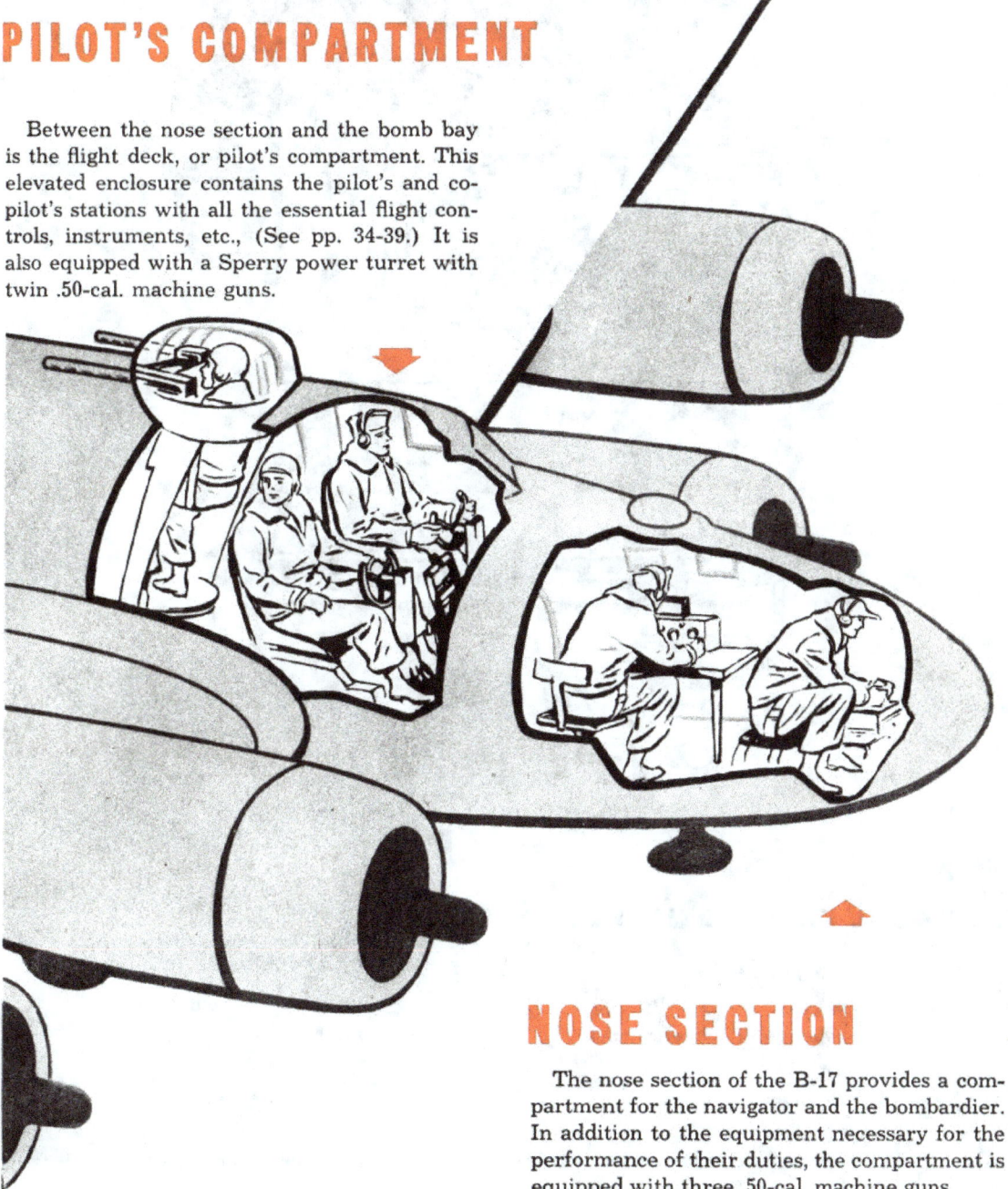

NOSE SECTION

The nose section of the B-17 provides a compartment for the navigator and the bombardier. In addition to the equipment necessary for the performance of their duties, the compartment is equipped with three .50-cal. machine guns.

RESTRICTED

PILOT'S OPERATIONAL EQUIPMENT

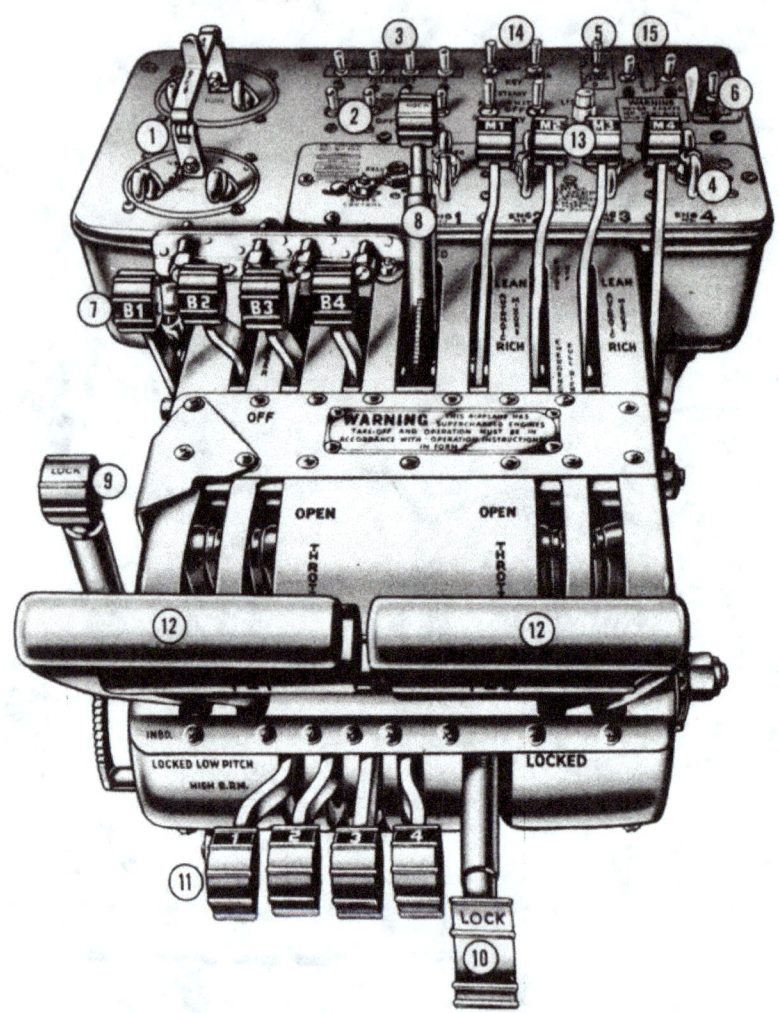

CONTROL PANEL AND PEDESTAL

1. Ignition switches
2. Fuel boost pump switches
3. Fuel shut-off valve switches
4. Cowl flap control valves
5. Landing gear switch
6. Wing flap switch
7. Turbo-supercharger controls (B-17F)
8. Turbo and mixture control lock
9. Throttle control lock
10. Propeller control lock
11. Propeller controls
12. Throttle controls
13. Mixture controls
14. Recognition light switches
15. Landing light switches

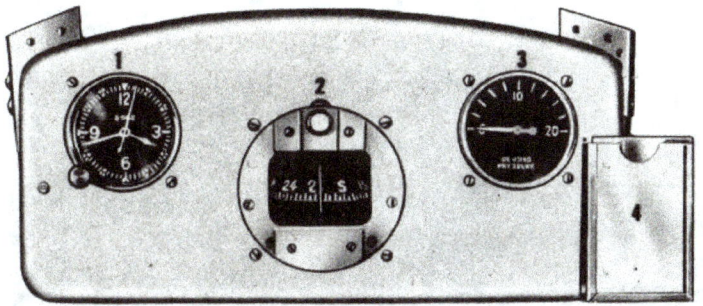

ABOVE WINDSHIELD

1. Clock
2. Compass
3. De-icer pressure gage
4. Compass card

LOWER CONTROL PEDESTAL

1. Elevator trim tab control
2. Automatic flight control panel
3. Rudder tab control
4. Elevator and rudder lock
5. Tailwheel lock

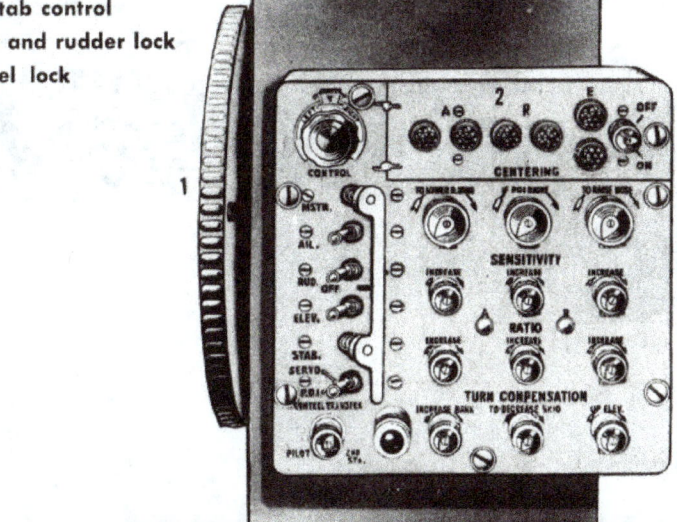

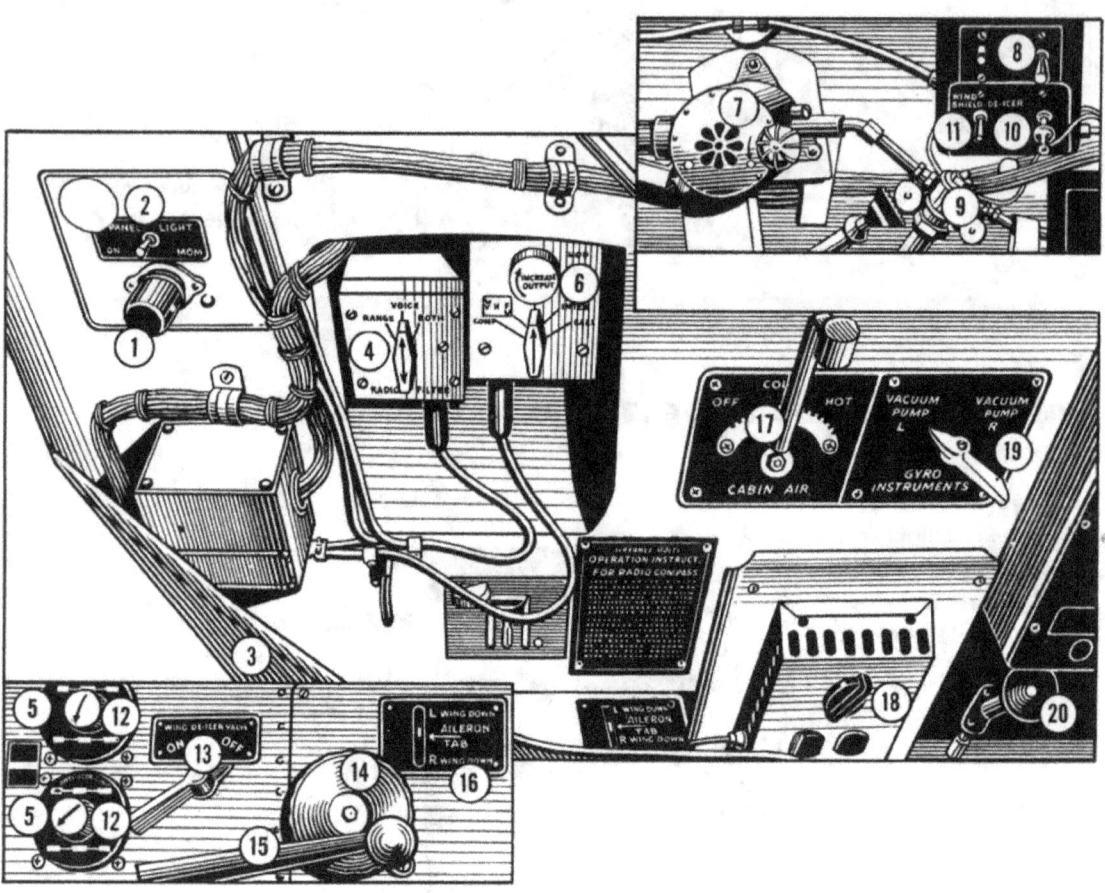

CONTROLS AT PILOT'S LEFT

1. Panel light
2. Panel light switch
3. Pilot's seat
4. Filter selector switch
5. Propeller anti-icer switch
6. Interphone jackbox
7. Oxygen regulator
8. Windshield wiper controls
9. Portable oxygen unit recharger
10. Windshield anti-icer switch
11. Windshield anti-icer flow control
12. Propeller anti-icer rheostats
13. Surface de-icer control
14. Aileron trim tab control
15. Pilot's seat adjustment lever
16. Aileron trim tab indicator
17. Cabin air control
18. Suit heater outlet
19. Vacuum selector valve
20. Emergency bomb release

PILOT'S CONTROL PANEL

1. Passing light switch
2. Running lights switch
3. Ammeters
4. Generator switches
5. Voltmeter
6. Battery switches
7. Alarm bell switch
8. Hydraulic pump servicing switch
9. Landing gear warning horn switch
10. Position lights switch
11. Voltmeter selector switch
12. Panel lights
13. Panel lights switch
14. Pitot heater switch
15. Interphone call light switch
16. Bomber call light switch
17. Inverter switch

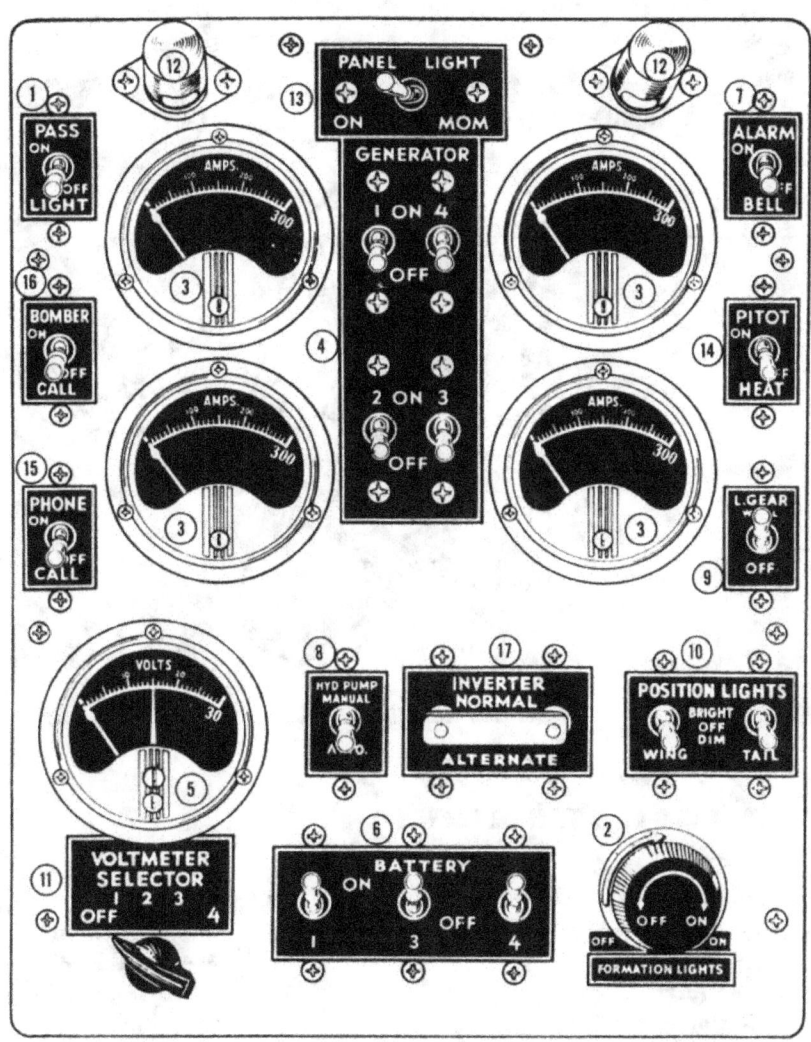

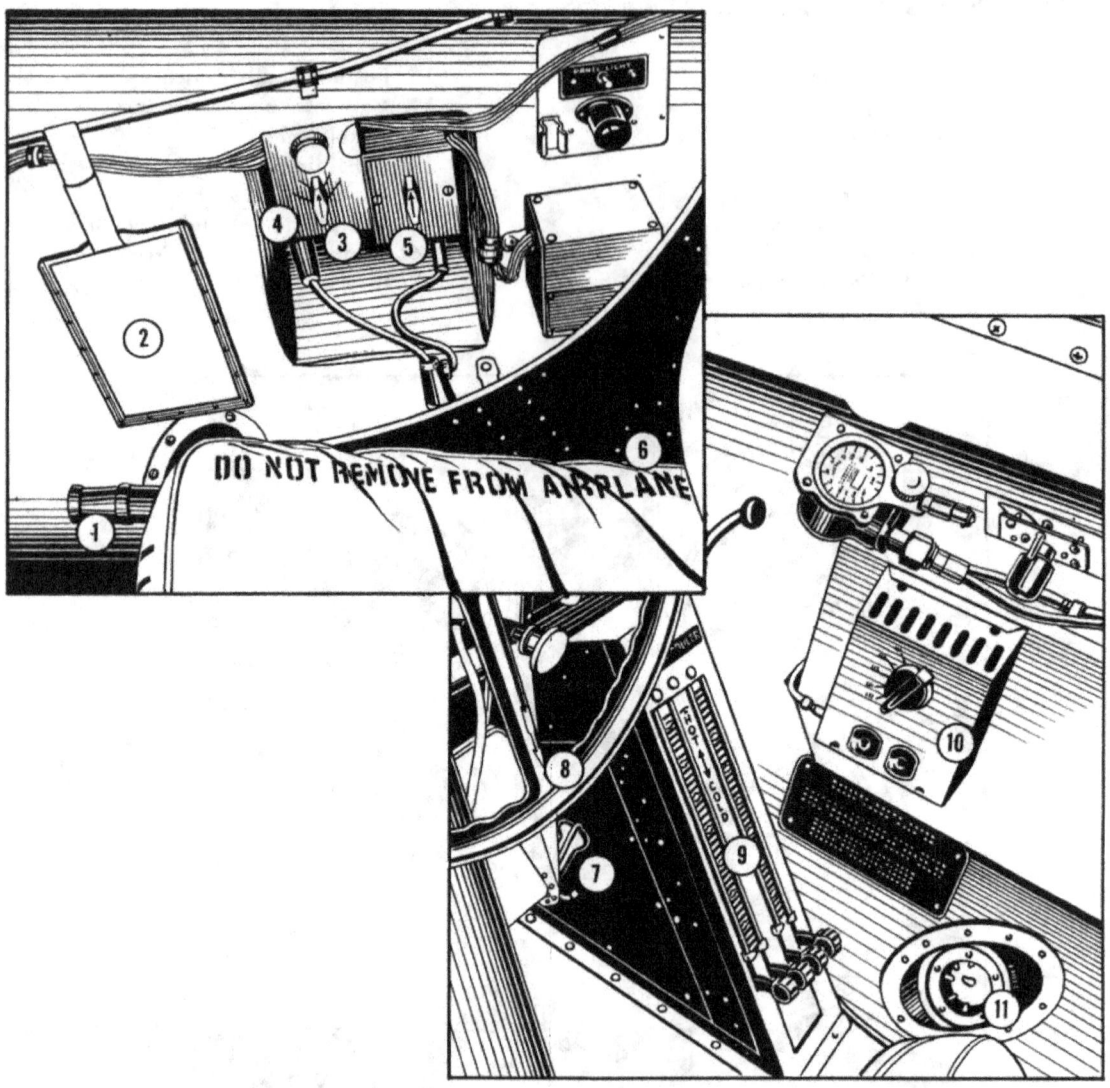

CONTROLS AT COPILOT'S RIGHT

1. Hydraulic hand pump
2. Checklist
3. Interphone selector switch
4. Interphone jackbox
5. Filter selector switch
6. Copilot's seat
7. Rudder pedal adjustment
8. Copilot's control wheel
9. Intercooler controls
10. Suit heater outlet
11. Engine primer

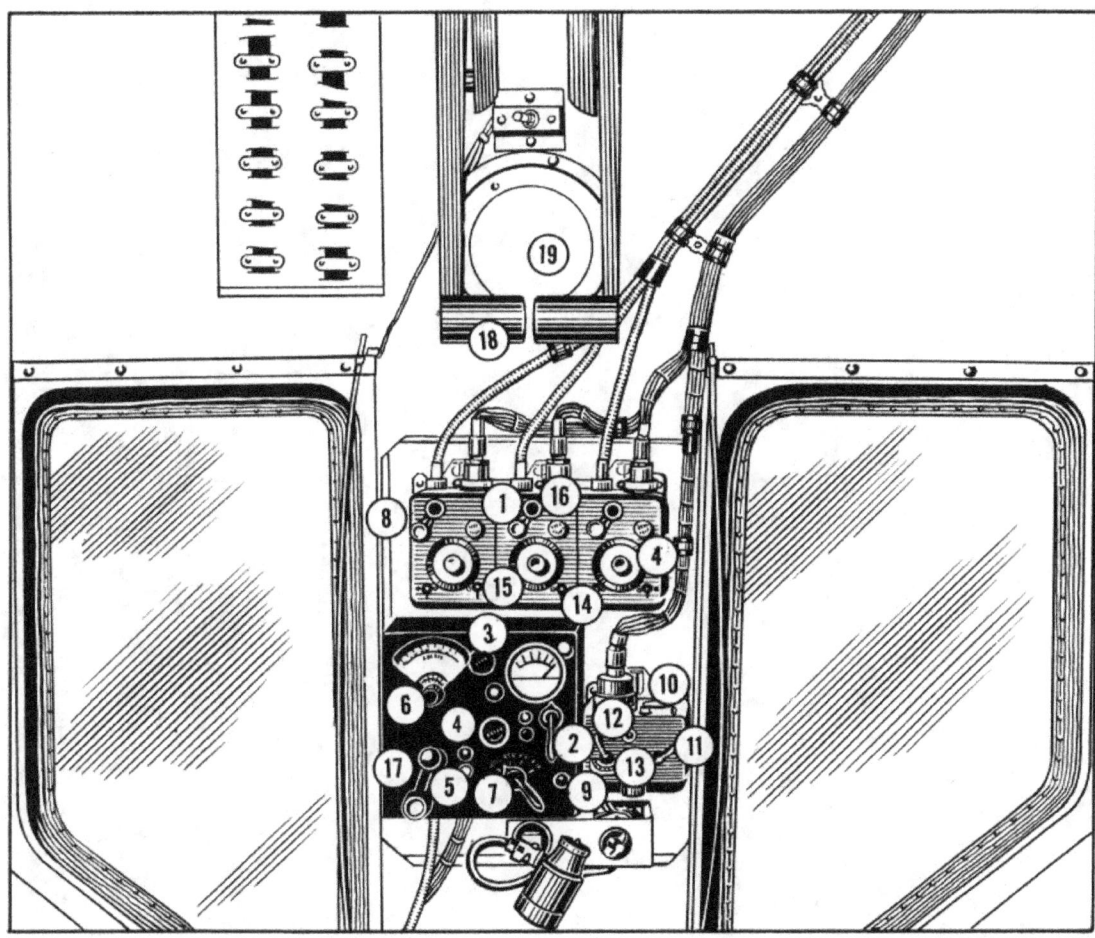

PILOT'S COMPARTMENT CEILING

1. Command receiver control unit
2. Loop control switch
3. Light control switch
4. Volume control
5. Control indicator lamp
6. Band selector knob
7. Power switch
8. Tuning crank
9. Control push button
10. Transmitting key
11. Transmission selector switch (Tone-CW-Voice)
12. Transmitter power switch
13. Channel selector switch
14. A-B channel switch
15. Signal selector switch
16. Volume control
17. Tuning crank
18. Emergency hand brake
19. Dome light

BOMB BAY

The bomb bay is aft of the pilot's compartment. Provision is made for releasable gasoline tanks in place of a bomb load. One tank may be carried on each side of the bomb truss. Tanks (or bombs) can be released electrically by the bombardier, or can be released by pulling one of the emergency release handles.

Bomb rack selector switches, installed on either side of the bomb bay, are used in conjunction with the rack selector switches on the bombardier's control panel. When either switch is "OFF" electrical release of bombs and fuel tanks is impossible.

A hand transfer pump is mounted on the aft bulkhead of the bomb bay and is used in case of the failure of the electric fuel pump.

"Tokyo tank" shut-off valves are mounted below the door at aft end of bomb bay. (In some installations these valves are in the radio compartment.)

A relief tube is located behind the dome light in the left bomb bay.

RADIO COMPARTMENT

The radio compartment is aft of the bomb bay section, and is reached from the flight deck by a catwalk through the bomb bay.

The radio compartment is equipped with one .50-cal. machine gun.

BALL TURRET

In the bottom of the waist section (aft of the radio compartment) provision is made for a Sperry ball-type power turret equipped with twin .50-cal. machine guns. This turret can be entered from within the airplane after takeoff.

TAIL GUNNER'S COMPARTMENT

The tail gunner's compartment is in the extreme end of the fuselage and is equipped with 2 direct-sighted .50-cal. machine guns. There are two ways of entering this compartment: (1) from the waist section through the tail-wheel compartment by means of a small door in the bulkhead, and (2) from the outside of the airplane through a small side door. The latter is an emergency exit and has an emergency release handle.

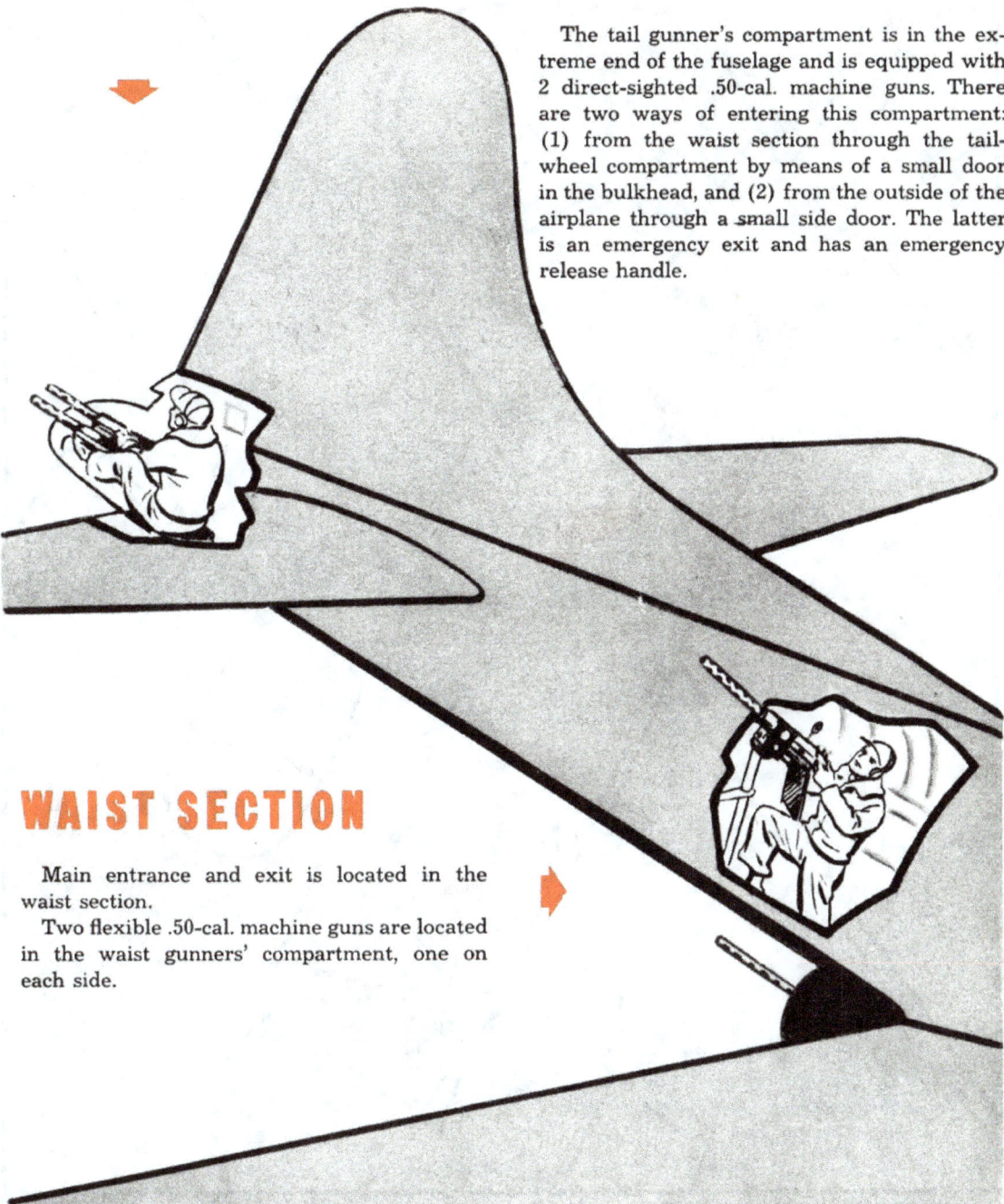

WAIST SECTION

Main entrance and exit is located in the waist section.

Two flexible .50-cal. machine guns are located in the waist gunners' compartment, one on each side.

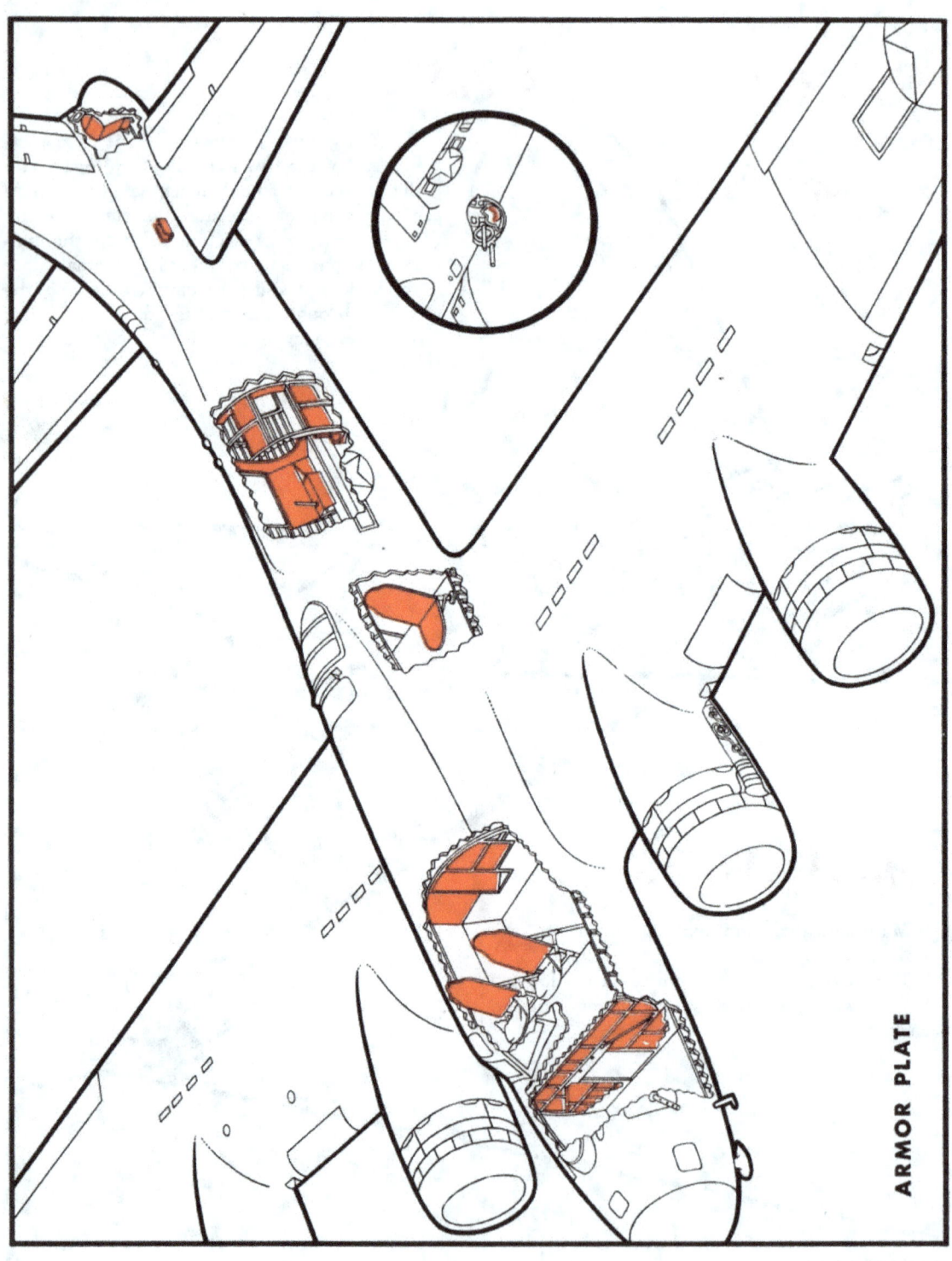

ARMOR

Protective armor plate mounted on rubber cushions is installed at crew stations throughout the airplane.

The pilot, copilot, and radio operator are protected by armor plate on the backs of their seats.

The bombardier-navigator compartment contains armor plate on the bulkhead at the rear of the compartment.

Armor protection for the top turret operator is placed on the aft side of the bulkhead at the rear of the pilot's compartment.

The gunner's seat in the ball turret is made of armor plate.

Armor plate in the waist gunner's compartment is installed above, below, and to the rear of each side window.

Padded armor plates and bulletproof glass protect the tail gunner.

The autopilot Servo motors above the tailwheel are protected by armor at the side and bottom.

Presenting the B-17G

The new B-17G, seventh major revision of the Flying Fortress, is now in operation at many bases in the continental United States. It incorporates a number of new features which have been developed as a result of the B-17's extensive combat experience.

The exterior appearance of the B-17G differs from that of the B-17F only in a few details.

Chin Turret

An electrically powered chin turret, equipped with two M-2 .50-cal. machine guns which have hydraulic charging mechanism, has been added.

The sights are synchronized with the turret turning mechanism, but they are not automatic computing sights.

The turret itself is located beneath the bombardier's station, and is operated by the bombardier.

The earlier models of the B-17G did not have cheek guns, but they are installed on later modifications on each side of the nose.

Pitot Static Mast

A single pitot static mast, placed just below the body center line aft of the chin turret and above the forward entrance hatch, replaces the two pitot static masts with which the B-17F is equipped.

Interior Changes

The navigator's compartment has been re-arranged. A larger table and a swivel chair are installed.

The flux gate compass and the radio compass have been relocated more conveniently on the bulkhead wall.

A shelf has been installed over the navigator's table.

A step has been added under the astrodome to facilitate the navigator's work in taking celestial shots.

Later B-17G's have the all electric bomb salvo system. Three toggle switches (on the bombardier's panel, above the copilot's instrument panel, and on the forward bomb bay bulkhead) allow emergency release of bombs. Any one switch opens the bomb doors electrically and releases the bombs. Entire operation takes about 12 seconds.

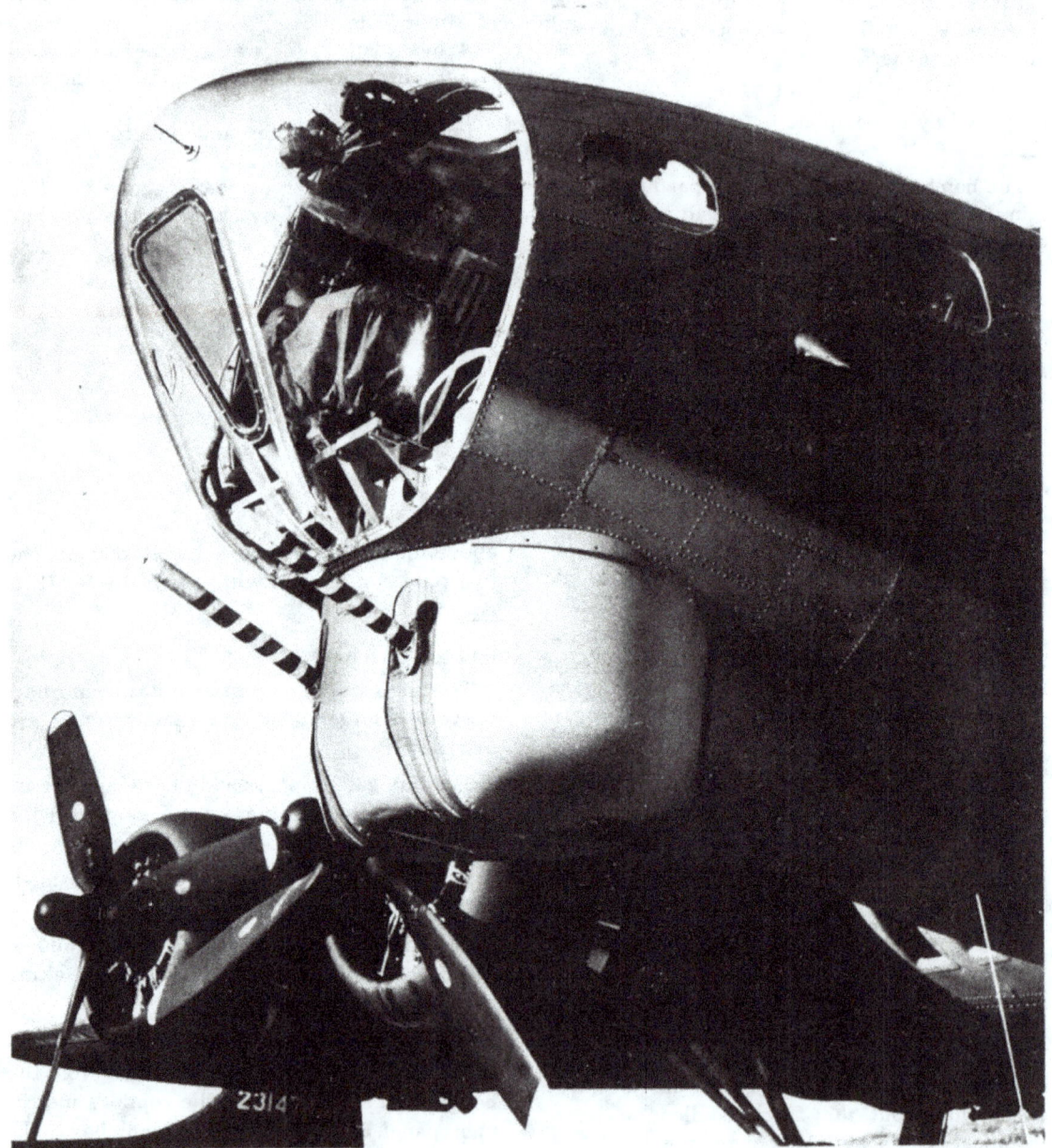

NEW CHIN TURRET ON B-17G

MISCELLANEOUS CHANGES

EQUIPMENT	B-17F	B-17G
Waist Windows	Removable	Fixed and staggered
Tachometer	Autosyn	Direct indication
Fuel Pressure Gages	Autosyn	Direct indication (liquid)
Oil Pressure Gages	Autosyn	Direct indication (liquid)
Manifold Pressure Gages	Autosyn	Direct indication (pressure)
Turbo-superchargers	B-2 (Manually Controlled)	B-22 (electronically controlled)
Turbo-superchargers	B-2 governed speed 23,400 r.p.m.	B-22 governed speed 26,400 r.p.m.
Tail Gunner's Compartment	Conventional; closed in by canvas cover on tail.	Enlarged; closed in completely by turret.
Windshield Knockout Panels	On some late modifications.	Installed.
Airspeed Indicator	Zero correction at low speeds. Too low indication at high speeds.	Zero correction at low speeds. Indicates too high at high speeds.
Booster Coil	Installed. Fires after top dead center.	Induction vibrator firing on top dead center. Requires change in starting procedure.
Emergency Oil Supply for Feathering	Not installed.	Oil tanks equipped with standpipe holding emergency supply.
Engine Fire Extinguisher System	Installed on few early models.	Installed on late airplanes.

INSPECTIONS *and Checks*

As a rated pilot with a certain number of hours in single engine and 2-engine aircraft to your credit, you are by now thoroughly indoctrinated in the vital importance of systematic and thorough inspections and checks.

That in itself is sufficient reason for you to stop now and reconsider the entire matter as you approach the B-17. For if the inspections and checks which you practiced as a pilot of single or 2-engine aircraft were important, they are doubly important now that you are the commander of one of the largest and most complex military airplanes in the world.

You are the commanding officer of an airplane costing approximately $250,000. You are responsible not only for the safety and efficiency of this valuable equipment but also for the lives of the crew.

Never take anything for granted about the airplane you are to fly. Not even the best preflight of the airplane by an unquestionably competent ground crew can relieve you of your responsibility to inspect personally the equipment you are about to take into the air. By now your own experience should tell you that **perfect maintenance** is almost an impossibility.

The responsibility is yours, and you can discharge the duties that it implies in one way only: **Check and double-check**.

Follow your routine of inspection scrupulously, and with your eyes wide open. Know what you are looking for and why.

Use your cockpit checklist. Use it properly, and at the indicated times: before starting the engines, before and after takeoff, before landing, on the final approach, after landing, etc.

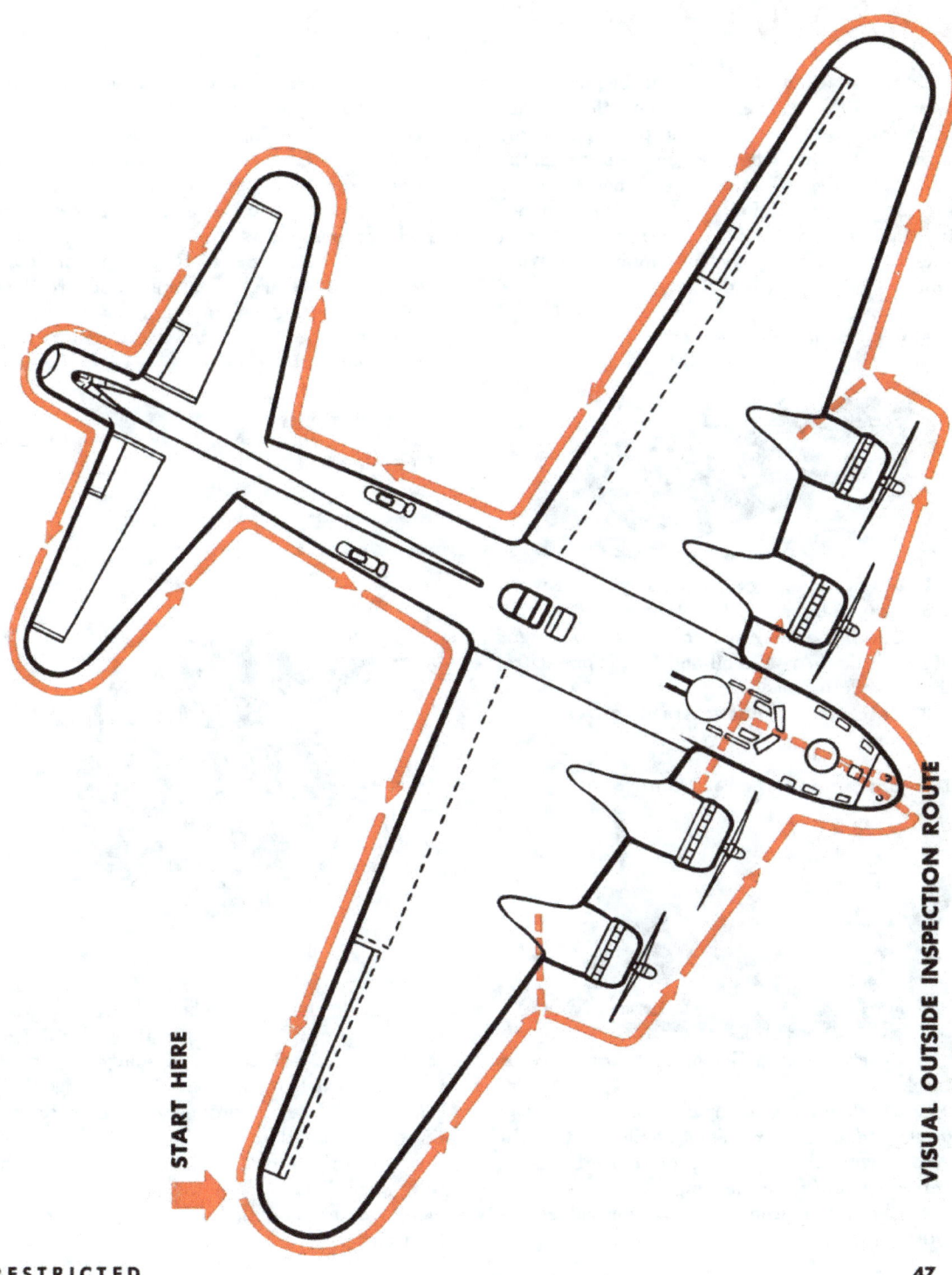

VISUAL OUTSIDE INSPECTION

Your visual outside inspection begins as you approach the airplane. Make a complete circuit of the airplane, beginning at the right wing, proceeding around the nose to the left wing, the ball turret, the tail surfaces, etc., before entering the waist door. Follow a definite sequence in making this outside inspection, checking each item in turn, always bearing in mind **what you are looking for** and **why**.

Beginning at the right wing

1. Check the **de-icer boots:** Any torn or worn spots, any roughness of contour?
2. Check the **wing center section:** Any signs of fuel leaks? Are the oil and fuel caps secure, the gaskets in place?
3. Check the **air ducts:** Are they free of obstructions?

That brings you to Power Plant No. 4

1. Check the **propeller blades:** Any nicks or cracks?
2. Check the propeller **anti-icer boots:** Look for looseness, for imbedded stones that might be thrown at the propeller, for signs of leaking anti-icing fluid from the **slinger ring.**
3. Check the propeller **governor cables** for tautness.

4. Check the **nacelle:** Look for loose fasteners or cowl flaps. Look for signs of oil leaks in the nacelle or on the engine. Look for dirt, stones, or other foreign matter wedged between the cylinders or the cylinder cooling fins.
5. Check the **engine exhaust systems** for cracks or loose joints.
6. Check **turbo wheels:** Revolve them slowly by hand to observe clearance and freedom. Look for missing buckets and for cracks between buckets. Be sure the waste gate is fully open; check it for proper looseness or freedom of movement.

Now follow the same procedure on Power Plant No. 3, checking each item as in the case of Power Plant No. 4.

Proceed to the right landing gear

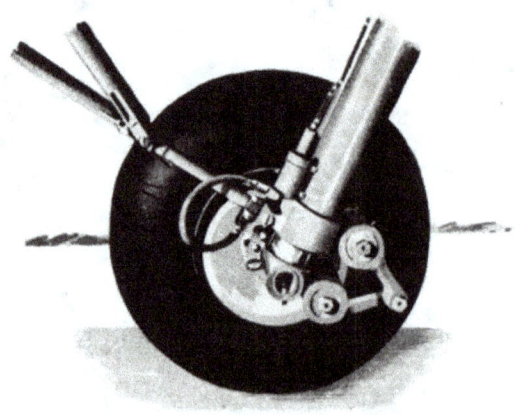

1. Check the **main wheel:** Look for worn spots on the tire, for cracks along the flanges in the rim. Check the tire visually for proper inflation; if it looks low, check with gage.
2. Check the conditions of the **hydraulic lines,** the condition and alignment of the **drag link** and **drag strut,** and the condition of the drag strut bolts. Check the joint between the oleo cylinder and axle knuckles for proper 1½-inch clearance.

3. Check the interior of the **wheel nacelle**, examining for play in the retracting screw, testing tautness of control cables and condition of pullies and electrical wiring. Look for excessive oil leaks throughout accessory section.

Inspect the nose of the airplane

1. Check **pitot tubes:** Have the covers been removed?
2. Check the **antennae:** Are they in proper place and with leads connected? Is the trailing antenna retracted?
3. Check the **marker beacon antenna** on the airplane's belly between the main entrance door and ball turret.

Continue your inspection of (1) the left landing gear, (2) Power Plant No. 2, (3) Power Plant No. 1, (4) the left wing, in each instance following the proper procedure as outlined.

Check ailerons and flaps

1. Check **aileron surfaces** and **trim tab alignment,** with controls in neutral. Apply pressure to the aileron to determine if controls are locked; check for excessive looseness.
2. See that **external locks** are removed.
3. Check the **flap** for alignment, and for holes or dents.

Inspect the ball turret

Be sure that it is in the locked position, that guns are stowed, and that the door is securely closed and locked.

Inspect the tail assembly

1. Check the **de-icer boots.**
2. Check the condition of the **elevators** and **rudder;** check the **trim tab** alignment. Be sure **external locks** have been removed.
3. Apply pressure to **control surfaces** to determine whether they are locked or free.
4. Check the **tail gun assembly:** Are the guns locked in position? Is the tail gunner's escape door closed?

Inspect the tailwheel assembly

1. Check the **tire:** For inflation, for cuts, for excessive wear.
2. Check **shear pin and slot:** Be sure they are not worn or rounded.

Finally, inspect your **right aileron and flap**—following the procedure used on the left aileron and flap.

You are now ready to enter the airplane.

RESTRICTED

INTERIOR VISUAL INSPECTION

CONTINUE YOUR SYSTEMATIC INSPECTION OF THE AIRPLANE AS YOU ENTER THE WAIST DOOR, AND BEFORE YOU PROCEED TOWARD THE FLIGHT DECK.

In the tail section

1. Check the **oleo** for the approximate clearance of 2⅝ inches.

2. Check the **drag link**, screw and entire assembly for alignment.

3. Examine the **control cables** for tightness.

4. Make certain that no baggage or equipment is in the tail section.

RESTRICTED

In the waist section

1. Check **guns** for proper stowage.
2. Make certain that **windows** are closed.
3. Check **control cables**. Are they too tight, or too loose? Are they free from coat hangers, magazines, newspapers, miscellaneous articles that may have become wedged in among them? Loose small equipment must not be stowed in the rear of the airplane. Violent action in rough air may throw such articles into the control cables.

In the radio compartment

1. Stop and check your weight and balance data in Form F-AN 01-1B-40. (See pp. 198-202: Weight and Balance.)

2. Check **Forms 1 and 1A.** Watch particularly for the symbol that may appear under the heading "Status Today." **A red diagonal** means that the airplane is flyable, but is not in perfect condition; **a red cross** means that there is a major defect in equipment and the airplane must not be flown; **a red dash** indicates that the required inspection has not been made.

When maintenance personnel place a red symbol under this heading, they are fulfilling their responsibility to you and to the safety of your flight. It's up to you to investigate the significance of the warning symbol, learn the nature of the trouble, and govern your flight accordingly.

The meaning of the **red diagonal** will be stated clearly on Form 1A. Be sure you understand the exact nature of the defect indicated.

3. Check the fuel and oil servicing section of Form 1A and the amounts serviced.

4. Pay particular attention to (a) the number of hours on each engine, (b) when the next inspection is due, (c) any notes that may have been entered by previous pilots or crew chiefs.

5. Check the **flight engineer's report** of preflight inspection. Discuss with the engineer any item that may indicate a questionable condition of the airplane or its equipment.

6. Make certain that the names of all crew members and passengers have been properly entered on the loading list. Sign the list and see that it is sent to Base Operations as required.

7. Ascertain that all on board are equipped with **parachutes,** that there is one extra parachute, and that this equipment is in proper condition. If an over-water flight is anticipated, make sure that all crew members have life vests.

8. Check **oxygen equipment:** The condition of masks, conditions of main oxygen system, the condition of walk-around bottles.

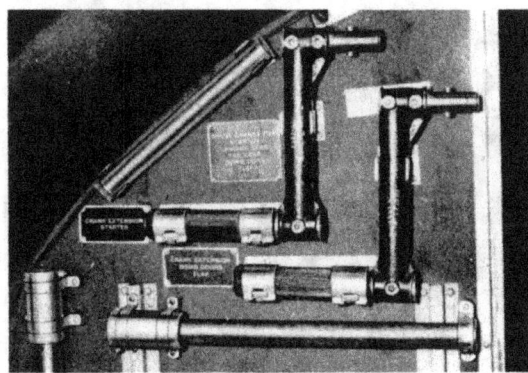

9. **Check the emergency landing gear hand crank:** Is it in proper place and locked?

10. Check **life raft emergency release handles** to be sure that they are properly set.

11. Check settings of the **command transmitter,** noting to what frequencies the transmitters are tuned.

12. Check Tokyo transfer valves for "CLOSED" position.

In the bomb bay section

1. Be sure the **bomb bay doors** are closed.
2. Check **bombs** or **racks** for proper installation.
3. Check the proper stowage of miscellaneous equipment.
4. If **bomb bay tanks** are installed, check the amount of fuel in each; be sure tank caps are properly secured, and rack selectors "OFF."

5. Check for excessive **gasoline fumes** in the bomb bay.

On the flight deck

1. Check the **upper turret:** switches "OFF," gun in aft position.
2. Check **fuel transfer valve and switch** in "OFF" position.
3. Examine floor and walls near the **hydraulic reservoir** for fluid leak. Check the **supply tank** for quantity of fluid.
4. Be sure that up-to-date copies of all required maps, radio facility charts, instrument let-down procedures, radio navigational aids, and direction finding charts are aboard.
5. See that sufficient **first-aid packets** are aboard.
6. Check the number, condition, and location of **fire extinguishers** aboard.
7. Have the ground crew pull the propellers through at least three revolutions to clear the combustion chambers of the engines, after making sure that all ignition and battery switches are "OFF."

You are now ready to begin actual preflight operations according to the cockpit checklist.

Cockpit Checklist

Every B-17 has a checklist on the copilot's side of the cockpit. Individual sections of the cockpit checklist are described at length in the chapters that follow.

Bear this in mind: **It is absolutely essential that the cockpit checklist be used properly by pilot and copilot at all times.**

The number of procedures necessary for the safe and efficient operation of the B-17 are far too many for even the most experienced pilot to carry in his head. The best trained pilots are likely to forget things occasionally. **There is no place for forgetfulness in flying the B-17!** Your cockpit checklist is the only sure safeguard against it.

Proper use of the checklist requires a **definite procedure** and **active cooperation** between the pilot and copilot.

1. The copilot takes the checklist in his hand and, in a clear, loud voice, calls out each item.
2. The specific operation or check is then performed, either by pilot or copilot (as specified by the checklist), whereupon pilot or copilot repeats aloud the item as "Checked!"

For example:

Copilot: **"Gear switch . . ."**

The pilot places his hand on the landing gear switch and ascertains that it is in the neutral position.

Pilot: **"Gear switch neutral."**

Copilot: **"Intercoolers . . ."**

The intercooler controls are on a separate stand to the right of the copilot. Therefore, the copilot places his hand on the controls and makes sure that they are in the "COLD" position.

Copilot: **"Intercoolers cold."**

There are some duties which must be performed by both the pilot and copilot, as in the case of checking the fire guard and calling "Clear!" before starting engines.

The copilot, with checklist in hand, has the responsibility of seeing that no item on it is left unchecked inadvertently. He must keep his finger on each item as it is called aloud, and not move on to the next item until he has personally seen the pilot check the first item or checked it himself.

Practical necessity demands that a few portions of the checklist (such as After Takeoff, After Landing, Running Takeoff, Go-Around, Approach, Before Takeoff) be memorized by pilot and copilot, since both will be too busy during these operations to refer to the printed checklist. In such cases, the checklist is called aloud from memory; but both pilot and copilot have the same responsibility to see that the checks and double-checks are made.

APPROVED B-17F and G CHECKLIST
REVISED 3-1-44

PILOT'S DUTIES IN RED
COPILOT'S DUTIES IN BLACK

BEFORE STARTING
1. Pilot's Preflight—COMPLETE
2. Form 1A—CHECKED
3. Controls and Seats—CHECKED
4. Fuel Transfer Valves & Switch—OFF
5. Intercoolers—Cold
6. Gyros—UNCAGED
7. Fuel Shut-off Switches—OPEN
8. Gear Switch—NEUTRAL
9. Cowl Flaps—Open Right—
   OPEN LEFT—Locked
10. Turbos—OFF
11. Idle cut-off—CHECKED
12. Throttles—CLOSED
13. High RPM—CHECKED
14. Autopilot—OFF
15. De-icers and Anti-icers, Wing and Prop—OFF
16. Cabin Heat—OFF
17. Generators—OFF

STARTING ENGINES
1. Fire Guard and Call Clear—LEFT Right
2. Master Switch—ON
3. Battery switches and inverters—ON & CHECKED
4. Parking Brakes—Hydraulic Check—On—CHECKED
5. Booster Pumps—Pressure—ON & CHECKED
6. Carburetor Filters—Open
7. Fuel Quantity—Gallons per tank
8. Start Engines: both magnetos on after one revolution
9. Flight Indicator & Vacuum Pressures CHECKED
10. Radio—On
11. Check Instruments—CHECKED
12. Crew Report
13. Radio Call & Altimeter—SET

ENGINE RUN-UP
1. Brakes—Locked
2. Trim Tabs—SET
3. Exercise Turbos and Props
4. Check Generators—CHECKED & OFF
5. Run up Engines

BEFORE TAKEOFF
1. Tailwheel—Locked
2. Gyro—Set
3. Generators—ON

AFTER TAKEOFF
1. Wheel—PILOT'S SIGNAL
2. Power Reduction
3. Cowl Flaps
4. Wheel Check—OK right—OK LEFT

BEFORE LANDING
1. Radio Call, Altimeter—SET
2. Crew Positions—OK
3. Autopilot—OFF
4. Booster Pumps—On
5. Mixture Controls—AUTO-RICH
6. Intercooler—Set
7. Carburetor Filters—Open
8. Wing De-icers—Off
9. Landing Gear
   a. Visual—Down Right—DOWN LEFT Tailwheel Down, Antenna in, Ball Turret Checked
   b. Light—OK
   c. Switch Off—Neutral
10. Hydraulic Pressure—OK Valve closed
11. RPM 2100—Set
12. Turbos—Set
13. Flaps ⅓–½ Down

FINAL APPROACH
14. Flaps—PILOT'S SIGNAL
15. RPM 2200—PILOT'S SIGNAL

RESTRICTED

Cockpit Checklist

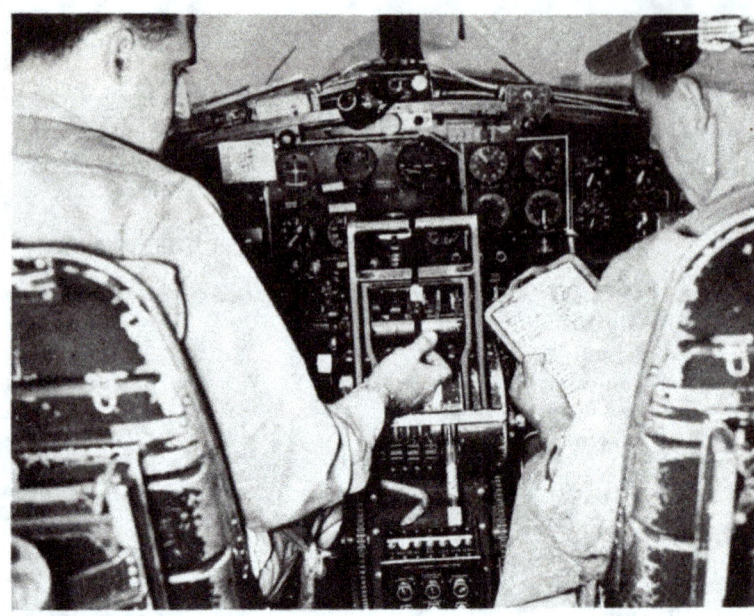

Every B-17 has a checklist on the copilot's side of the cockpit. Individual sections of the cockpit checklist are described at length in the chapters that follow.

Bear this in mind: **It is absolutely essential that the cockpit checklist be used properly by pilot and copilot at all times.**

The number of procedures necessary for the safe and efficient operation of the B-17 are far too many for even the most experienced pilot to carry in his head. The best trained pilots are likely to forget things occasionally. **There is no place for forgetfulness in flying the B-17!** Your cockpit checklist is the only sure safeguard against it.

Proper use of the checklist requires a **definite procedure** and **active cooperation** between the pilot and copilot.

1. The copilot takes the checklist in his hand and, in a clear, loud voice, calls out each item.

2. The specific operation or check is then performed, either by pilot or copilot (as specified by the checklist), whereupon pilot or copilot repeats aloud the item as "Checked!"

For example:

Copilot: **"Gear switch . . ."**

The pilot places his hand on the landing gear switch and ascertains that it is in the neutral position.

Pilot: **"Gear switch neutral."**

Copilot: **"Intercoolers . . ."**

The intercooler controls are on a separate stand to the right of the copilot. Therefore, the copilot places his hand on the controls and makes sure that they are in the "COLD" position.

Copilot: **"Intercoolers cold."**

There are some duties which must be performed by both the pilot and copilot, as in the case of checking the fire guard and calling "Clear!" before starting engines.

The copilot, with checklist in hand, has the responsibility of seeing that no item on it is left unchecked inadvertently. He must keep his finger on each item as it is called aloud, and not move on to the next item until he has personally seen the pilot check the first item or checked it himself.

Practical necessity demands that a few portions of the checklist (such as After Takeoff, After Landing, Running Takeoff, Go-Around, Approach, Before Takeoff) be memorized by pilot and copilot, since both will be too busy during these operations to refer to the printed checklist. In such cases, the checklist is called aloud from memory; but both pilot and copilot have the same responsibility to see that the checks and double-checks are made.

APPROVED B-17F and G CHECKLIST
REVISED 3-1-44

PILOT'S DUTIES IN RED
COPILOT'S DUTIES IN BLACK

BEFORE STARTING
1. Pilot's Preflight—COMPLETE
2. Form 1A—CHECKED
3. Controls and Seats—CHECKED
4. Fuel Transfer Valves & Switch—OFF
5. Intercoolers—Cold
6. Gyros—UNCAGED
7. Fuel Shut-off Switches—OPEN
8. Gear Switch—NEUTRAL
9. Cowl Flaps—Open Right—
   OPEN LEFT—Locked
10. Turbos—OFF
11. Idle cut-off—CHECKED
12. Throttles—CLOSED
13. High RPM—CHECKED
14. Autopilot—OFF
15. De-icers and Anti-icers, Wing and Prop—OFF
16. Cabin Heat—OFF
17. Generators—OFF

STARTING ENGINES
1. Fire Guard and Call Clear—LEFT Right
2. Master Switch—ON
3. Battery switches and inverters—ON & CHECKED
4. Parking Brakes—Hydraulic Check—On-CHECKED
5. Booster Pumps—Pressure—ON & CHECKED
6. Carburetor Filters—Open
7. Fuel Quantity—Gallons per tank
8. Start Engines: both magnetos on after one revolution
9. Flight Indicator & Vacuum Pressures CHECKED
10. Radio—On
11. Check Instruments—CHECKED
12. Crew Report
13. Radio Call & Altimeter—SET

ENGINE RUN-UP
1. Brakes—Locked
2. Trim Tabs—SET
3. Exercise Turbos and Props
4. Check Generators—CHECKED & OFF
5. Run up Engines

BEFORE TAKEOFF
1. Tailwheel—Locked
2. Gyro—Set
3. Generators—ON

AFTER TAKEOFF
1. Wheel—PILOT'S SIGNAL
2. Power Reduction
3. Cowl Flaps
4. Wheel Check—OK right—OK LEFT

BEFORE LANDING
1. Radio Call, Altimeter—SET
2. Crew Positions—OK
3. Autopilot—OFF
4. Booster Pumps—On
5. Mixture Controls—AUTO-RICH
6. Intercooler—Set
7. Carburetor Filters—Open
8. Wing De-icers—Off
9. Landing Gear
   a. Visual—Down Right—DOWN LEFT
      Tailwheel Down, Antenna in, Ball Turret Checked
   b. Light—OK
   c. Switch Off—Neutral
10. Hydraulic Pressure—OK Valve closed
11. RPM 2100—Set
12. Turbos—Set
13. Flaps ⅓—½ Down

FINAL APPROACH
14. Flaps—PILOT'S SIGNAL
15. RPM 2200—PILOT'S SIGNAL

AFTER LANDING
1. Hydraulic Pressure—OK
2. Cowl Flaps—Open and Locked
3. Turbos—Off
4. Booster Pumps—Off
5. Wing Flaps—Up
6. Tailwheel—Unlocked
7. Generators—OFF

END OF MISSION
1. Engines—Cut
2. Radio—On ramp
3. Switches—OFF
4. Chocks
5. Controls—LOCKED
6. Form 1

GO-AROUND
1. High RPM & Power—High RPM
2. Wing Flaps—Coming Up
3. Power reduction
4. Wheel Check—OK Right—OK LEFT

RUNNING TAKEOFF
1. Wing Flaps—Coming Up
2. Power
3. Wheel Check—OK Right—OK LEFT

SUBSEQUENT TAKEOFF
1. Trim Tabs—SET
2. Wing Flaps—UP
3. Cowl Flaps—Open Right—OPEN LEFT
4. High RPM—CHECKED
5. Fuel—Gals per tank
6. Booster Pumps—ON
7. Turbos—SET
8. Flight Controls—UNLOCKED
9. Radio Call

SUBSEQUENT LANDING
1. Landing Gear
   a. Visual—Down Right—DOWN LEFT
      Tailwheel Down, Ball Turret
      Checked
   b. Light—ON
2. Hydraulic Pressure—OK
3. RPM 2100—Set
4. Turbo Controls—Set
5. Wing Flaps ⅓–⅓ Down
6. Radio Call

FINAL APPROACH
7. Flaps—PILOT'S SIGNAL
8. RPM 2200—PILOT'S SIGNAL

FEATHERING
1. Throttle Back
2. Feather
3. Mixture and Fuel Booster—Off
4. Turbo Off
5. Prop Low RPM
6. Ignition Off
7. Generator Off
8. Fuel Valve Off

UNFEATHERING
1. Fuel Valve On
2. Ignition On
3. Prop Low RPM
4. Throttle Cracked
5. Supercharger Off
6. Unfeather
7. Mixture Auto-Rich
8. Warm up Engine
9. Generator On

SEQUENCE OF POWER CHANGES

INCREASING POWER
1. Mixture Controls
2. Propellers
3. Throttles
4. Superchargers

DECREASING POWER
1. Superchargers
2. Throttles
3. Propellers
4. Mixture Controls

STARTING

Your cockpit checklist becomes effective just as soon as you have stepped across the flight deck and climbed into your seat:

Pilot's Preflight

The duties under this heading—which began with your outside visual inspection of the airplane and continued as you passed through the interior from the rear section to the flight deck —have been completed.

Form 1A

Form 1A was checked, completed, and signed after you inspected the radio compartment.

Controls and Seats

Check your controls—rudder, elevators, and ailerons. Put them through their full range of operation to insure freedom of movement and proper direction of operation.

Now both pilot and copilot adjust their seats, rudder pedals, and safety belts to insure freedom of movement and control through the full range of operation. Proper adjustment of these items is particularly important when the use of full rudder becomes necessary.

Fuel Transfer Valves and Switch

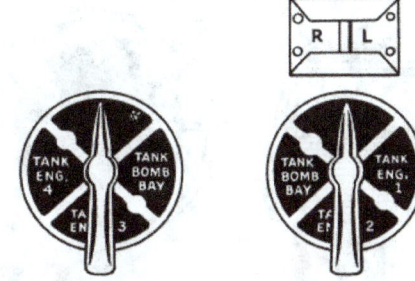

Check your fuel transfer valves and switch to be sure they are in the "OFF" position. Remember: if they are not turned "OFF," you may pump one of the engine tanks dry, and waste a lot of fuel from the overflow of the tank into which the gasoline is being pumped.

Intercoolers

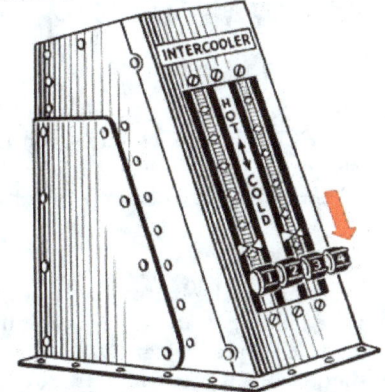

The copilot checks the intercooler controls and ascertains that they are in the "COLD" position. (The function of the intercoolers in connection with the operation of the superchargers is explained on pp. 169-173.)

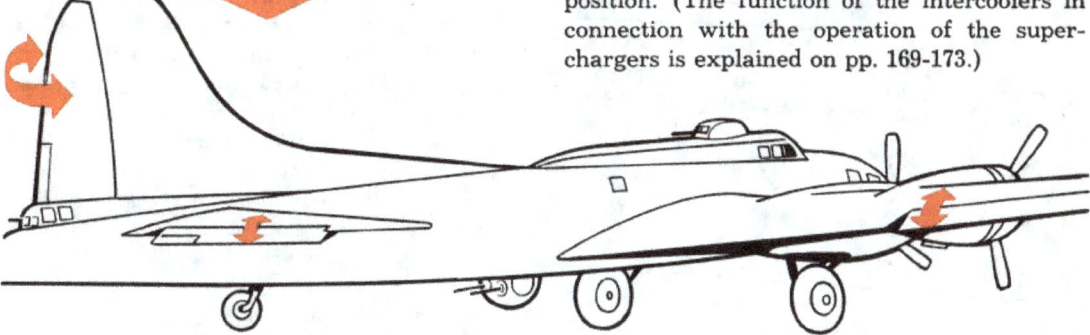

Gyros

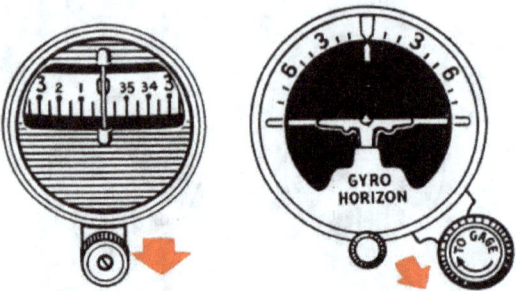

Check your gyro instruments to be sure they are uncaged.

Fuel Shut-off Switches

Check the fuel shut-off switches to be sure they are "OPEN." These switches control the fuel supply to the engines. They should be left open at all times except in emergencies.

Landing Gear Switch

Before turning on the battery switches, make sure that the landing gear toggle switch has not been turned "UP" inadvertently. **Landing gear switch should be at neutral and the switch guard in position.**

Cowl Flaps

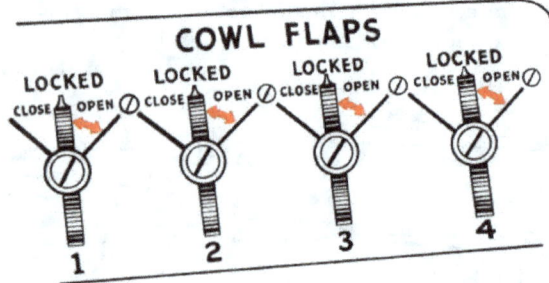

Regardless of outside air temperature, the cowl flaps must be open before starting engines to avoid spot heating. Pilot and copilot check: "Cowl flaps open left"; "cowl flaps open right!"—"Locked." Cowl flaps should be in the "LOCKED" or neutral position to prevent creeping or loss of pressure.

Turbo-superchargers

Turbos are always turned "OFF" during starting. With the supercharger on, the waste gate is closed. A backfire could blow out the waste gate or damage the supercharger.

Idle Cut-off

Check to insure that mixture controls are in the "IDLE CUT-OFF" position.

Throttles Closed

Close the throttles, then move them forward to the setting for approximately 1000-2000 rpm. Engines will start much easier with the throttles in this position. (**After the engine has been started** and begins to run smoothly, bring the throttles back to approximately 800-1000 rpm. The throttles should not be moved backward and forward in an attempt to smooth out the engine. This results in a lean mixture, backfiring, and increased fire hazard.

High RPM

Place the propeller controls in "HIGH RPM" and adjust the lock to hold securely.

Automatic Pilot

Place the automatic pilot switches in the "OFF" position and leave them there until after takeoff. Takeoffs with the automatic pilot on have resulted in accidents. Autopilot pressure is supposed to be low enough so that it can be overpowered by the manual controls, but on takeoff the busy pilot probably will be slow to recognize this condition and apply sufficient pressure on the controls quickly enough. So, before starting, check: **Autopilot—"OFF."**

De-icers and Anti-icers

Place the wing de-icer control valve, and the propeller anti-icer knobs and control switch in

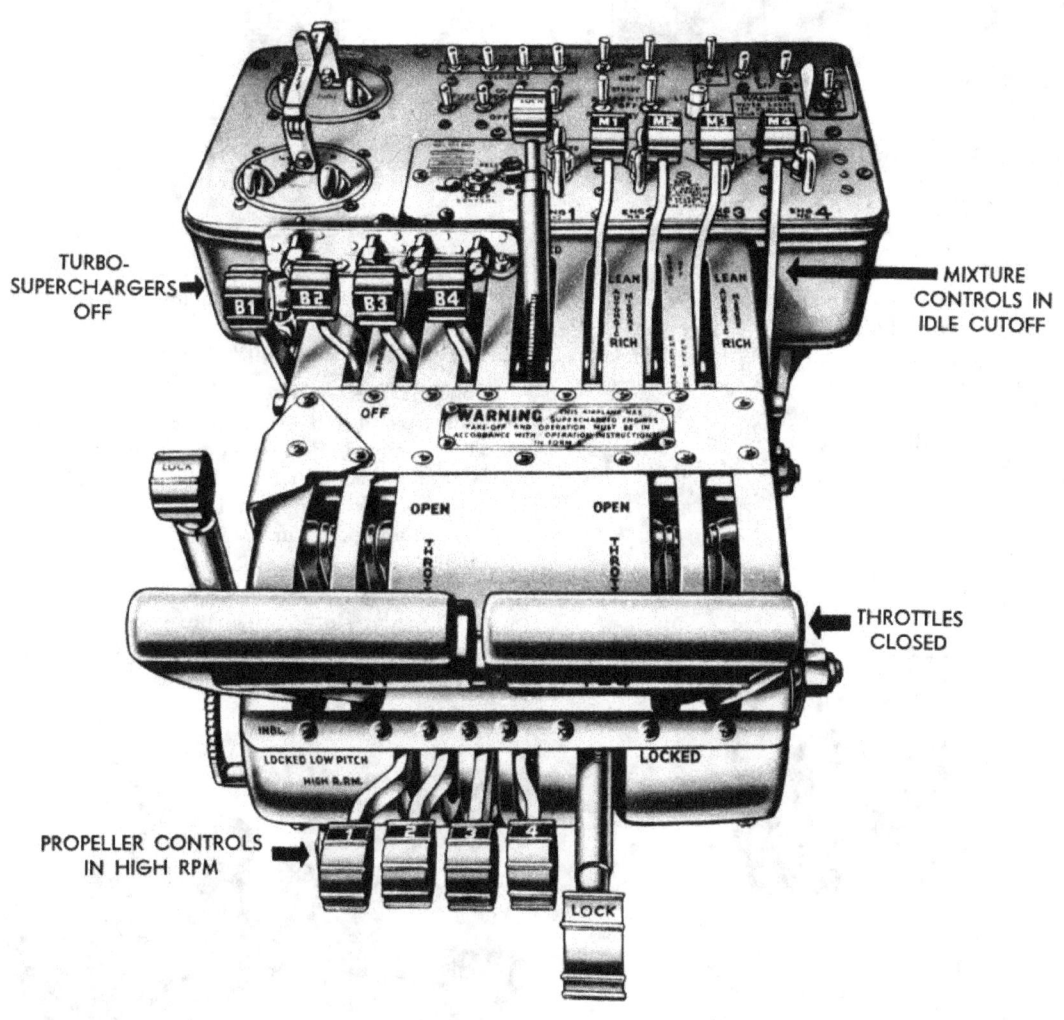

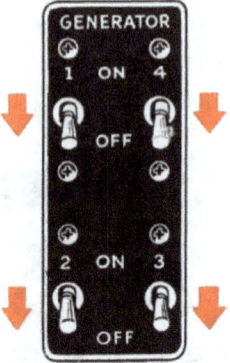

the "OFF" position. Since the action of the wing de-icer boots disturbs the flow of air over the lifting surfaces and materially increases stalling speed, the wing de-icers are never used on take-off. The propeller anti-icing fluid is not needed on takeoff since ice is unlikely to form quickly enough. (When flights are to be made into icing conditions, both these systems should be checked thoroughly prior to takeoff.

Cabin Heat

Put the cabin heat control in the "OFF" position and keep it there during all ground operations. This will allow an unrestricted flow of air through the heating system radiator, and tend to prevent boiling of the fluid. Use the cabin heater only in the air.

Generators

Keep the generator switches "OFF" until the airplane is in the takeoff position with engines running up to 1500 rpm. This prevents closing the points of the generator cut-out relays and the consequent reverse flow of current while taxiing. Don't use generators below 1500 rpm.

Check Fire Guard and Call "Clear"

Look out the window and be sure that the fire guard is posted at his proper station—behind and to the right of the engine being started.

The starting sequence is engines No. 1, No. 2, No. 3, No. 4. This sequence should be followed in order to avoid confusion of the ground crew.

The pilot calls "Clear left," and the copilot calls "Clear right," before engines are started on either side. Both will make sure that the mechanic hears the call, and signifies (by voice or by hand signal) that all is clear.

Master and Ignition Switches

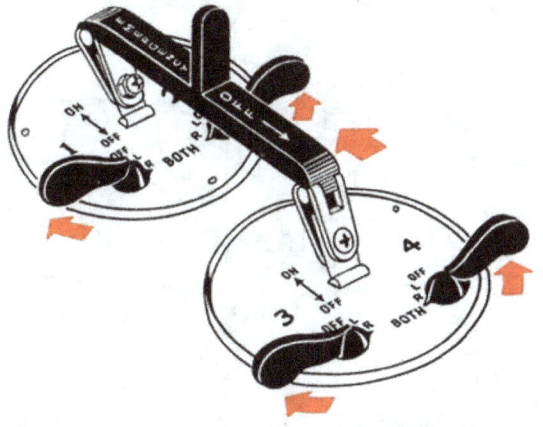

Place the bar switch in the "ON" position. Put all ignition switches in "BOTH" position. (Note: Except in the B-17G where individual ignition switches are turned "ON" **after the corresponding engine is meshed** and the propeller has turned through one revolution.)

Battery Switches and Inverter

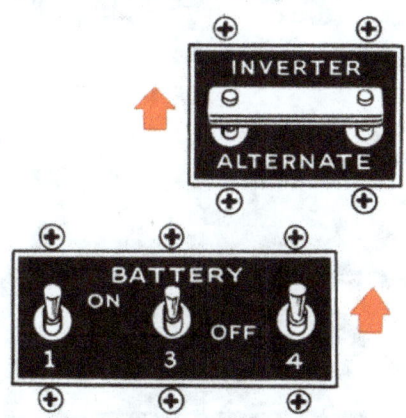

Move the inverter switch to "NORMAL." Then operate each battery switch separately to detect a battery in need of charging. Check the fuse and solenoid. Return all 3 battery switches to "ON." Now check inverter in "ALTERNATE" position. Return the inverter switch to "NORMAL," and leave it there during flight. The alternate inverter is used only in the event that the normal inverter fails. The alternate remains new and unused for such emergency.

Parking Brakes and Hydraulic Check

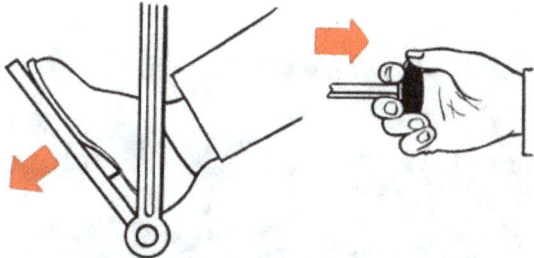

Copilot sets and locks the parking brakes.
Check the pressure gages for sufficient hydraulic pressure (600-800 lb.). Check the switch on the pilot's switch panel for the "AUTO" or "ON" position—depending on the type of switch installed. If the emergency pressure system is low, recharge by opening the manual shut-off (star) valve. This will build up pressure in both systems to approximately 800 lb. After recharging, close the manual shut-off (star) valve. If emergency system is installed, operate levers to insure that upon application pressure does not drop immediately to zero. Be sure that the selector is in "NORMAL" position, and that the reservoir is filled with hydraulic fluid.

Booster Pumps

Turn on the booster pumps and check to see that each gives from 6 to 8 lb. pressure. The fuel booster pump is an independent electrically driven source of extra fuel pressure. It takes the place of the wobble pump for both starting and emergencies, and augments the engine-driven fuel pump at high altitudes. As a safety measure, it is always turned on for takeoff and landing, for flights below 1000 feet, and for flights above 10,000 feet.

Carburetor Filters

Carburetor air filters must be "ON" ("OPEN") for engine starting and all operations up to 8,000 feet in the B-17F (15,000 feet in the B-17G). Check amber warning light for "ON."

In dust conditions filters may be left "ON" in the B-17F up to 15,000 feet (20,000 feet in the B-17G).

But under no circumstances should the carburetor air filters be left "ON" above these limits. When intake air passes through the carburetor air filters at such altitude the turbo-superchargers must speed up to maintain desired manifold pressure. This can result in turbo overspeeding.

Fuel Quantity

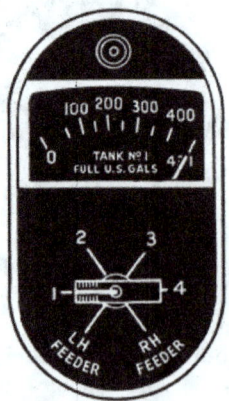

Check the fuel gages for quantity of fuel in each tank. Remember that the fuel gages are electric and will not operate unless the battery switches and inverter are on.

Start Engines

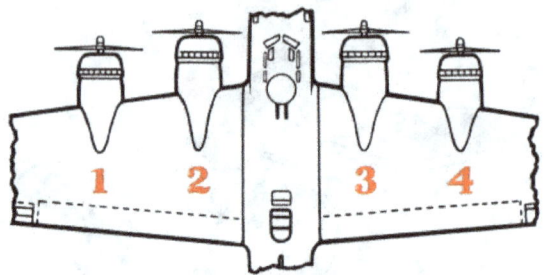

1. The sequence of starting engines is: No. 1, No. 2, No. 3, and No. 4.
2. Be sure the engine being started has been pulled through 3 or 4 complete revolutions.

3. If fire extinguisher system is installed, set the selector switch to the engine being started.

4. Indicate to the ground crew (by holding up fingers) which engine is being started.
5. When the copilot is ready, he will notify the pilot: "Starting by to start No. 1."

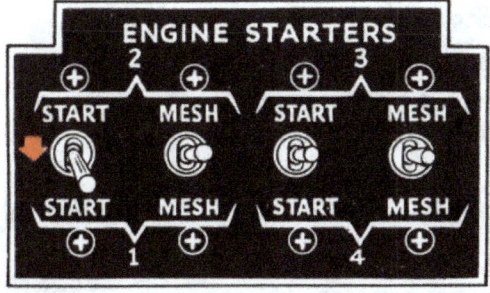

6. Direct the copilot: "Start No. 1." The copilot will then energize Engine No. 1, and at the same time expel all air from the primer with the number of strokes necessary to obtain a solid fuel charge.. The primer must be **held down** until needed again.

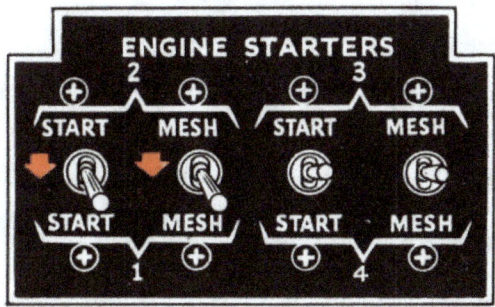

7. After approximately 12 seconds of energizing, direct the copilot to "Mesh No. 1." The copilot, while still holding the starting switch at "START," moves mesh switch to the "MESH" position. At the same time he primes with strong, steady strokes until the engine fires.

8. If the engine fails to fire after the starter has turned it over 4 or 5 times, the copilot must release both switches quickly while the propeller is still turning. This prevents damage to, or sticking of, the starter. If the starter dog sticks and the engine turns over while re-energizing, stop re-energizing immediately, cut the ignition switch, and release the starter dog by turning the propeller in the direction of rotation.

9. When the engine fires, move the mixture control to "AUTO-RICH" immediately.

10. If the engine stops, return the mixture control to "OFF" immediately, and repeat the starting procedure. As soon as engine is running, copilot calls: "Oil pressure." Pilot notes pressure, and responds: "Coming up" when pressure reaches 50 lb. sq. in.

11. If no oil pressure is indicated within 30 seconds after starting, stop the engine and determine the cause.

12. Warm up engines at 1000 rpm until oil temperature of 40°C is indicated.

13. If it is necessary to engage by hand, signal to the ground crew by raising a clenched fist and pulling down an imaginary starter handle. One of the ground crew will pull the handle on the nacelle. Meanwhile, hold down both the starter and the mesh switches in the "ON" positions. The booster coil will function only when the mesh switch is on.

14. Repeat the same starting procedure on No. 2, No. 3, and No. 4 engines, in that order.

Flight Indicator and Vacuum Pressures

When an engine that operates a vacuum pump (No. 2 and No. 3 Engines) is started, check the rapid response of the flight indicator. With vacuum pump operating, the flight indicator should erect itself within a few moments. Sluggish response at this time indicates poor operation of the instrument. At the same time check (1) vacuum pressure – approximately 3.75" to 4.25"; and (2) both pumps for proper operation.

(**Note:** If Jack and Heintz flight indicator is installed, it must be erected with the caging knob.)

Radio

Set the switches on the command receiver to proper positions. Turn the transmitter switch "ON." Set the selector to the desired transmitting frequency. Turn volume controls on jackboxes to maximum output. Set the selector switch on filter box to "VOICE," and selector switch on the jackbox to "COMMAND."

Check Instruments

Check instruments for
(1) proper operation
(2) readings within the proper range

Oil pressure
DESIRED .... 75 LB.
MAXIMUM .... 80 LB.
MINIMUM .... 70 LB.

Oil temperature
DESIRED .... 70°C.
MAXIMUM .... 88°C.
MINIMUM .... 60°C.

Cylinder-head temperature
DESIRED .... 170°C.
MAXIMUM .... 205°C.
MINIMUM .... 125°C.

Fuel pressure
DESIRED .... 12-16 LB.

Carburetor air temperature
DESIRED .... 15°C.
MAXIMUM .... 38°C.

Free air temperature
DOES GAGE REGISTER APPROXIMATE OUTSIDE TEMPERATURE?

Tachometers
STEADY INDICATION

Manifold pressures
STEADY INDICATION

Hydraulic pressures
DESIRED .... 600-800 LB.

Clock
CHECK AND SET

Magnetic compass
IS THE FLOAT LEVEL?

Flap position indicator
CHECK FOR OPERATION

Check Lights

Check all warning lights. (The fuel gage warning light is tested by pushing in the light bulb.)

If the flight is to extend after dark, check all other lights for proper functioning: landing, passing, wingtip, fluorescent, compartment, radio compass, and identification lights. A flashlight, in good working order, should be carried.

Check fuse panel covers for adequate supply of extra fuses.

Check the warning bell for proper operation.

Crew Report

Check the crew to make sure that all doors and hatches are closed, and that all crew members are at proper stations, **with headsets on.**

Radio Call and Altimeter Setting

Call the tower for clearance, and obtain altimeter setting. Set and check the altimeter. If the setting varies more than 75 feet from field elevation, ask for another check.

TAXIING

There is only one reason for a taxiing accident: carelessness. The pilot who taxies slowly and observes the few basic rules will never have the inexcusable experience of damaging an airplane in simple ground operation.

The pilot experienced on heavier types of aircraft should understand the reasons for taxiing slowly. Primarily they are safety considerations, and the mechanical limitations of the brakes.

Safety considerations are so obvious that they need little explanation. The pilot who taxies slowly **always has control of the airplane** and can stop whenever and wherever he chooses.

The mechanical limitations of brakes make slow taxiing mandatory. You can't stop 50,000 lb. of fast-moving airplane in a short space. It takes tremendous frictional energy to slow down and stop this large mass. Moreover, frequent application of brakes, which is necessary when the airplane is not taxied slowly, causes excessively high brake temperature and eventual brake failure.

1. Before wheel chocks are removed, check hydraulic pressure: it should be 600 to 800 lb.

2. Taxi from the parking area **with all 4 engines running**, using the outboard engines for turning. Keep your inboard engines idling at not less than 500 rpm, with just enough friction lock applied to prevent the throttles from creeping. Don't lock the throttles of the inboard engines tightly; you may need them in an emergency.

3. Never taxi faster than a ground crew man can walk.

4. Use brakes only to slow down or stop the airplane, or to aid in making turns, when necessary. At all other times, keep your feet off the brake pedals with your heels on the floor. Even slight pressure will result in brake heating. When it becomes necessary to use brakes, slide your feet up on the pedals until the balls of the feet are squarely on the brake controls. Apply brakes smoothly and firmly. (Don't pat the brakes.) As soon as the airplane is under control, release brakes and return heels to floor.

5. For all straight ahead taxiing—even for a short distance—keep the tailwheel locked.

6. Before making a turn, have the copilot unlock the tailwheel. Make turn by using the throttles, with as little brakes as possible.

7. Always make turns with the inside wheel rolling. Pivoting on the inside wheel causes excessive wear on the tire and places a heavy torque strain on the gear.

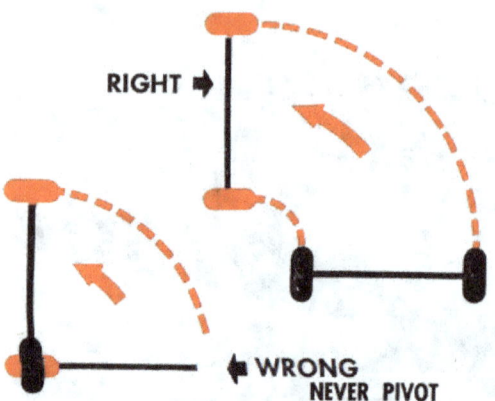

RIGHT →

← WRONG NEVER PIVOT

8. If a side wind blows the airplane off a straight line, wait until you reach the other side of the runway, then unlock the tailwheel and redirect the airplane, crabbing away from the windward side of the runway in a series of arcs or S's. (See cut.) Use the outboard engine on the side from which the wind is blowing to decrease the rapidity of your drift toward the windward side of the runway.

9. Hold the aileron and elevator controls in a neutral position, so that these control surfaces will be streamlined with wing surfaces and elevator stabilizers respectively. Don't try to taxi an airplane by steering with the control wheel as you would drive a car.

10. Take particular care never to allow the inboard engines to idle slowly enough to load up. During any one period of parking, don't permit them to idle at less than 1000 rpm. If you have to taxi over a long distance, stop and run up the engines high enough and often enough to keep them clear.

11. Don't try to taxi if hydraulic pressure is low and will not build up. (You will only lose what little pressure you have.) Have the airplane towed back to the line.

12. Have your auxiliary power unit turned on for all ground operations. This insures operation of the electrically operated hydraulic pump.

Remember that cold weather and low rpm do not work together. Therefore, when the temperature is low clear the engines oftener than usual. Naturally, this will require an increased use of brakes.

UNLOCK TAILWHEEL AND REDIRECT USING OUTBOARD ENGINE

WIND

TAKEOFF TECHNIQUE

Taxi to run-up area, park into the wind when possible, and call for engine run-up check. Co-pilot responds: "Brakes set." Make sure that the throttles are set at not less than 1000 rpm.

Trim Tabs

Set the trim tabs for takeoff. Check to see that all 3 tabs are at the "0" (zero) setting. Incorrect setting of any trim tab on takeoff can cause a serious accident, especially if the airplane is heavily loaded.

Exercise Turbos and Propellers

Advance throttles to 1500 rpm, and run the turbo controls through their range several times. Still maintaining 1500 rpm, and with turbo controls "OFF," run the propellers through to "LOW RPM," then back to full "HIGH RPM."

Allow ample time for the propellers to change pitch. Watch carefully for the drop in rpm (approximately 300-400 rpm) indicated by the tachometers.

When rpm decreases to approximately 1100, return the propeller controls to "HIGH RPM." At the same time, return turbo controls to the "OFF" position.

Repeat these turbo and propeller exercises three or four times, or more if the outside air temperature is below 0°C.

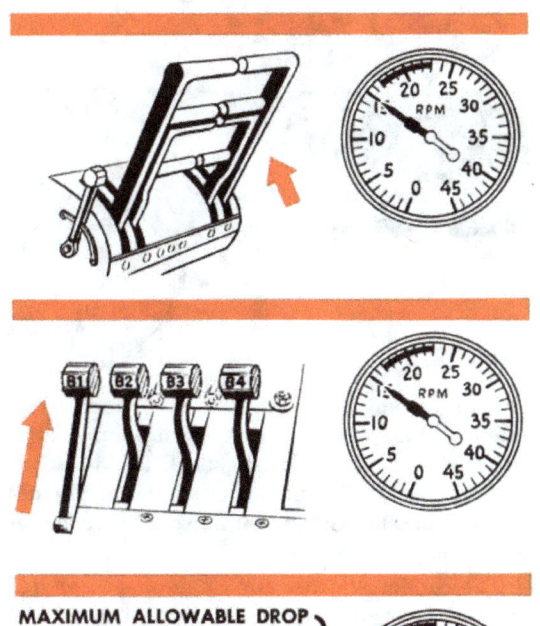

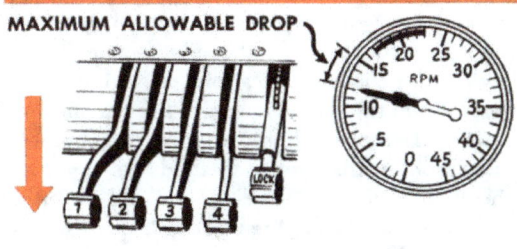

Check Generators

Check the generators while the engines are operating at 1500 rpm. Check them for ample output; and, by using the voltmeter selectors, check for voltage output.

With all generators on, check the pitot heaters by watching for a rise in the ammeter reading. Then turn the pitot heater off.

Turn generators "OFF." Idle engines at not less than 1000 rpm.

Run Up Engines

Run up engines one at a time and in sequence. Open throttle to 28" manifold pressure. Then turn to left magneto, back to both, then to right magneto, then back to both. **Do not operate on one magneto for more than 5 seconds at a time.**

The copilot watches for roughness of engine operation by observing any drop in rpm. The pilot keeps an eye on the engine nacelle and cowling for visible indications of engine roughness. While the visual check of the nacelles and cowling is more reliable than the tachometer indication, utilize **both** methods as a double check. If much roughness is noticed on either magneto, run the engine up to full throttle **with turbo off** for about 10 seconds; then return to 28" manifold pressure, and check again.

During the pilot's ignition check, the copilot will check the following items:

Fuel pressure
DESIRED . . . . 12 TO 16 LB. SQ. IN.
MAXIMUM . . . . 16 LB. SQ. IN.
MINIMUM . . . . 12 LB. SQ. IN.

Oil pressure
DESIRED . . . . 75 LB. SQ. IN.
MAXIMUM . . . . 80 LB. SQ. IN.
MINIMUM . . . . 70 LB. SQ. IN.

Oil temperature
DESIRED . . . . 70°C.
MAXIMUM . . . . 88°C.
MINIMUM . . . . 60°C.

Cylinder head temperature
MAXIMUM . . . . 205°C.

Run-up Procedure

1. After checking magnetos, hold at 28" Hg., and move turbo control full forward against the stop.

2. Wait for increased manifold pressure (usually about 5-8" Hg. surge). This indicates that turbo wheel is turning up to speed.

3. Run throttle forward, and adjust turbo to give desired takeoff setting.

Remember that because of direct linkage control, the waste gate will open immediately when turbo control is moved toward closed position, and will lag when moved forward. Therefore, care should be exercised in adjusting control so that excessive full throttle operation is avoided on the ground.

Check rpm. Normally, for ground operation, rpm can vary between 2400 and 2500 maximum.

Reduce throttle to 1000 rpm.

Repeat the foregoing run-up procedure on engines No. 2, No. 3 and No. 4 in sequence.

Throttle Technique

The most comfortable and effective way to handle the throttles of the B-17 for operation of all 4 engines is to hold the **right hand palm upward,** thus grasping all 4 throttle handles firmly within the palm and fingers. (See cut.)

Holding them in this manner permits an *easy* wrist movement for **progressively** leading and controlling the throttles, and tends to favor the inboard throttles.

Progressively leading the throttles means alternately advancing right and left engines—in other words, **walking the throttles** steadily forward.

Adjustment of the throttle friction lock should be just enough to prevent the throttles from creeping. Don't jam the lock lever hard forward; you'll only have to struggle to loosen the lock each time you want to change throttle settings. Friction should be such that (1) throttle creeping is prevented, and (2) the throttle can be moved without too much pressure in case of emergency.

Before Takeoff

After engine run-up has been completed, make your radio call to the tower and request permission to taxi to takeoff position. Do not taxi on the runway until this radio contact has been completed. Bear in mind that it may be necessary for the tower to respond by using a red or green Aldis light.

Pilot and copilot should check visually to be sure the runway is clear and that no aircraft are landing. The tower is not infallible.

When cleared by the tower, instruct the copilot to unlock brakes. Then, with engines idling at not less than 800-1000 rpm, taxi on to the runway. Take a position that will allow use of the full runway. See that all windows are closed and locked. Cowl flaps must be left open on takeoff. Call for takeoff check.

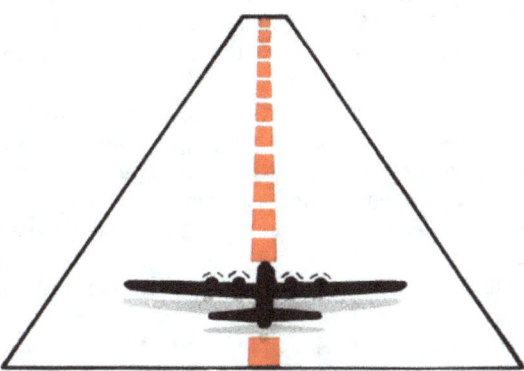

See that the airplane is lined up properly with the runway. Instruct the copilot to "Lock tailwheel." The copilot will lock the tailwheel as the airplane is slowly rolling forward, and will inform you: "Tailwheel locked; light out—Gyros."

Check the gyros. Set the directional gyro to correspond with the magnetic compass. When lined up for takeoff, check your compass reading with the runway heading. Pilot responds: "Gyros set."

Copilot calls: "Generators" as throttles are advanced for takeoff. When 1500 rpm is reached, pilot turns on generators with left hand.

TAKEOFF

Duties of the pilot, copilot, and flight engineer on takeoff are well defined. Each has specific duties to perform, and it is important that all three should have an over-all understanding of the takeoff procedure.

1. Apply power gradually, **progressively leading the throttles.** (See p. 69.) Avoid **overcontrol,** which will require reduction of power on either side.
2. Keep your right hand on the throttles.
3. During the takeoff run, maintain directional control with rudder and throttles. **Keep ailerons neutral.**
4. Always take off from a 2-point, tail-low attitude. (The 3-point takeoff should never be attempted except in an emergency.) Don't attempt to pull the airplane into the air. Normally when you have attained an airspeed of approximately 110-115 mph, moderate back pressure on the control column will enable the airplane to fly itself off the ground.
5. The copilot follows through on the throttles, keeping his left hand in position to make adjustments for variations in manifold pressure, and prepared to take immediate action in such emergencies as runway propellers or overspeeding turbos.
6. The copilot's principal duty on takeoff is to watch the engine instruments, particularly manifold pressure, rpm, pressure gages, and temperature gages. He must divide his attention between engine instruments and the actual progress of the takeoff.
7. Takeoff distances for various field conditions and airplane loading are stated specifically on the seat-back operating instructions and in AN 01-20EF-1 and AN 01-20EG-1.
8. After the airplane has left the ground, and you are positive that you have sufficient flying speed and that everything is under control, signal to the copilot to raise the landing gear. The copilot will apply brakes gently to stop the rotation of the wheels, and raise the gear. Both pilot and copilot make a visual check, and acknowledge the retraction of the main wheels (Pilot: "Landing gear up left." Copilot: "Landing gear up right." The flight engineer checks and reports "Tailwheel up.") The copilot places the landing gear switch in the neutral position.
9. The B-17 is so constructed that very little change in trim will be required after takeoff.
10. Depending upon elevation and gross load, signal the copilot either to reduce or shut off the turbos.
11. Reduce power upon attaining an airspeed of 140 mph. To obtain normal climb attitude, the pilot reduces the throttles to a manifold pressure between 32" and 35" Hg. in the transition type B-17, and 35" Hg. in the normally operated tactical airplane. Then the copilot reduces rpm to 2300.
12. The copilot will make the necessary adjustments of cowl flaps to regulate cylinder-head temperature during the climb. They should be closed whenever possible.

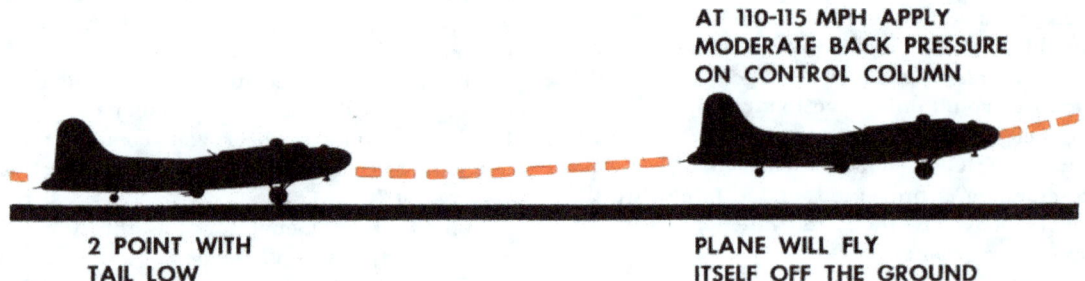

2 POINT WITH TAIL LOW

AT 110-115 MPH APPLY MODERATE BACK PRESSURE ON CONTROL COLUMN

PLANE WILL FLY ITSELF OFF THE GROUND

RUNNING TAKEOFF

This type of takeoff does not vary much in basic technique from the normal takeoff.

1. Make a normal 3-point landing.
2. When the airplane has settled into the landing roll, inform the copilot: "Running takeoff."
3. The copilot immediately checks propeller controls for "HIGH RPM," and places the flap switch in the "UP" position.
4. Now apply power, walking up throttles steadily and smoothly. Avoid abrupt throttle movement.
5. Use rudder for directional control. The airplane still has most of its landing speed when power is applied. If directional control is difficult before full power is attained, use coordinated throttle and rudder.

From this point forward, the operation is the same as a normal takeoff. Complete the usual after-takeoff check: (1) Signal copilot for "Wheels up" if leaving traffic; (2) reduce power; (3) adjust cowl flaps; (4) make check of "Wheels up right." "Wheels up left." Engineer: "Tailwheel up."

CROSSWIND TAKEOFF

The crosswind takeoff requires **use of more rudder** and more **differential throttling** than the normal takeoff.

Most modern airfields are so constructed that there is seldom any occasion for taking off in an extreme crosswind. However, because the large vertical surfaces of the airplane are exposed to any wind from the side the airplane will tend to veer **into** the wind. Therefore, the technique of the crosswind takeoff is extremely important and frequently useful.

Remember that the important elements in the crosswind takeoff control, in order of importance, are: (1) rudder, (2) differential throttling, and (3) the downwind brake **only as a last resort.**

Use rudder to keep the airplane straight as long as possible. However, in a strong crosswind, if use of rudder is not sufficient to keep the airplane straight, apply more power to the upwind engines. Remember that progressive application of power (on all 4 engines) is necessary to attain takeoff speed as quickly as possible.

If the upwind engines have been used all the way to the stop and the rudder still will not straighten the airplane, **only then apply slight reduction of power on the downwind engines.** Under most crosswind conditions, this should not be necessary.

Don't attempt to use the downwind brake except as a last resort.

CORRECT WITH RUDDER AND UPWIND ENGINES

RESTRICTED

CLIMBING AND CRUISING

The rate at which an airplane will climb is obtained directly from the difference between the **power required for level flight** and the **power available** from the engines. This difference is the **reserve power** which can be used for climbing.

Climbing the B-17

Flight tests have shown that for B-17's of all weights, the difference between **power required** for level flight and **power available** reaches a maximum at approximately 135 mph IAS. For stability purposes, another 5 mph is added as a safety margin. Therefore, **make your climb at 140 mph IAS,** except on instruments.

Climbing on Instruments

On instruments below 20,000 feet, climb at 150 mph IAS. Here again an allowance has been made in the recommended airspeed for a safety margin.

Power Settings for Climbing

Power settings for the normal climbing conditions are as follows:

	RPM	MANIFOLD PRESS.	MIXTURE
GRADE 100 FUEL			
Maximum climb	2300	38" Hg.	Auto-Rich
Desired climb	2300	35" Hg.	Auto-Rich
GRADE 91 FUEL			
Maximum climb	2300	37" Hg.	Auto-Rich
Desired climb	2300	35" Hg.	Auto-Rich
Desired climb (light transition planes)	2300	32–35" Hg.	Auto-Rich

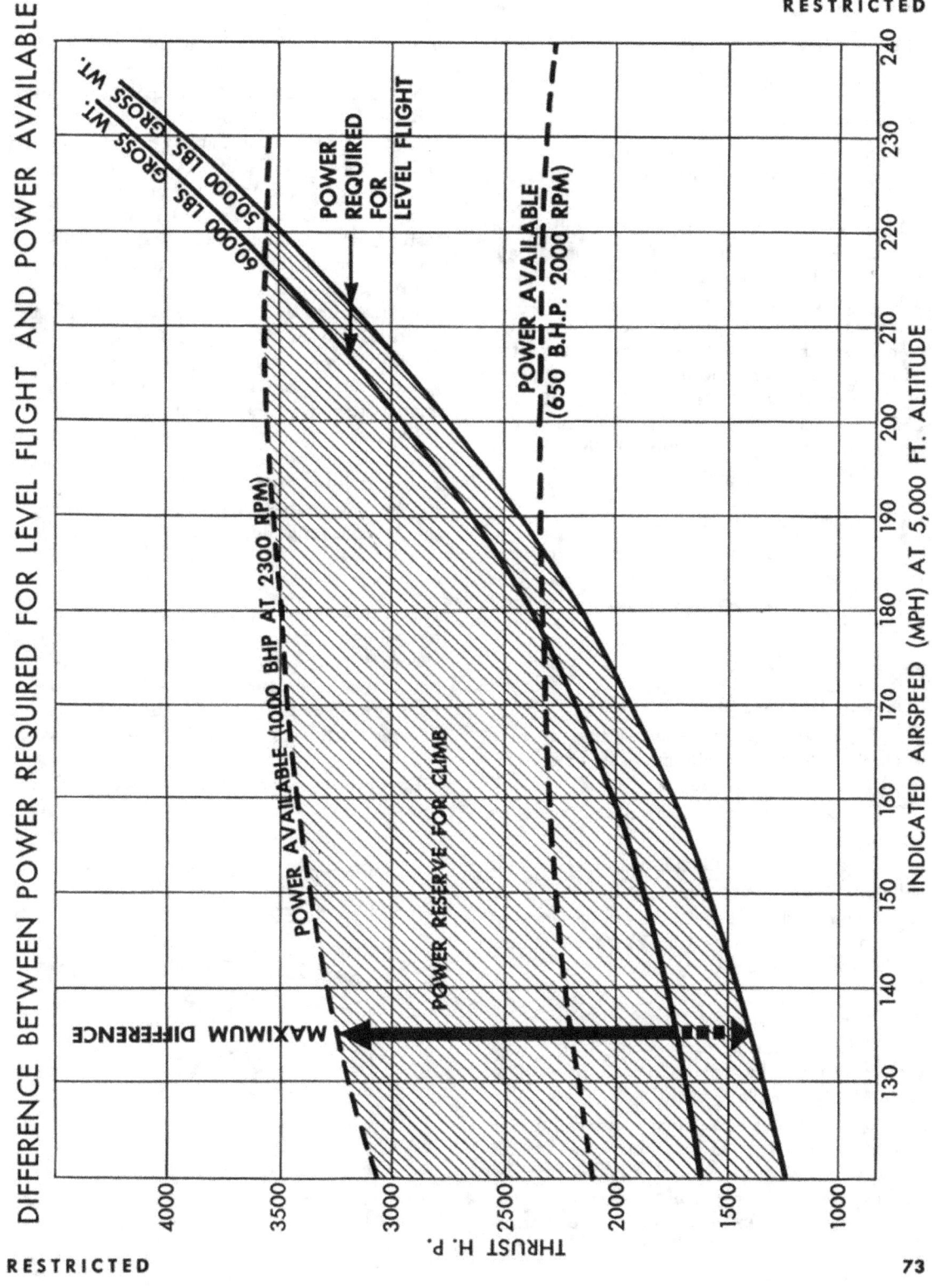

Angle of Climb

The proper angle of climb should be judged by airspeed, obstacles to be cleared, and the attitude of airplane. Trim the airplane to relieve control pressures, and synchronize propellers as soon as climbing power settings are established.

In B-17's with full crews, and with guns and turrets installed, with or without bomb load, 35" Hg. and 2300 rpm at 140 mph will give a desired attitude and rate of climb. However, in a transition airplane from which this equipment has been removed, this power setting for climbing may cause the airplane to assume a high climb attitude while maintaining 140 mph IAS. Take these things into consideration, and remember: the important thing is to maintain normal climbing attitude and airspeed.

Auto-rich for All Climbs

For all climbs leave mixture controls in auto-rich. At high power the proportion of fuel to air must be relatively high to assist in cooling and prevent detonation.

Effects of Increasing Altitude

As altitude increases:

1. Engines get hotter the longer they operate at climbing power, thereby increasing cylinder-head and oil temperatures.
2. IAS gradually falls; atmospheric pressure gradually decreases.
3. It becomes more difficult for man to obtain sufficient oxygen from the atmosphere.

Remember these conditions which develop with increasing altitude. Consider their effects on (a) your airplane, (b) your crew.

Engine Heat

1. **Cylinder-head Temperatures.** Adjust cowl flaps to maintain head temperatures just below the maximum of 205°C.
2. **Use of Cowl Flaps.** Keep in mind that the position of cowl flaps affects your rate of climb because of added drag and disturbance of the airflow. However, do not hesitate to use them to keep the cylinder-head temperatures within operating limits. Use the minimum setting that will maintain the temperatures desired.
3. **Oil Temperatures.** Oil temperatures can be reduced more quickly by decreasing the engine rpm and manifold pressure than by reducing the throttles alone. Another way to reduce both cylinder-head and oil temperatures is to shallow your climb so that your IAS is 5 to 10 mph faster than normal climbing airspeed. This will not cause much loss in your rate of climb.

In case of high cylinder-head and oil temperatures, you can use **emergency** (full) rich mixture. This will dissipate the heat rapidly, but will also cause loss of power and excessive gas consumption. Therefore, use it only long enough to reduce temperatures.

Excessive temperatures are often caused by failure of the automatic feature of the carburetor, thereby producing too lean a mixture. Placing the control in emergency rich corrects this by enriching the mixture.

Decreasing Air Temperature

1. **Carburetor Air Temperature.** On an extended climb, check constantly to be sure your carburetor air temperature is either above or below the icing range: from $-5°C$ to $+15°C$. Particularly if the humidity is high, you can develop carburetor ice with little or no warning.

Carburetor temperatures above 38°C are likely to cause detonation. Control your carbu-

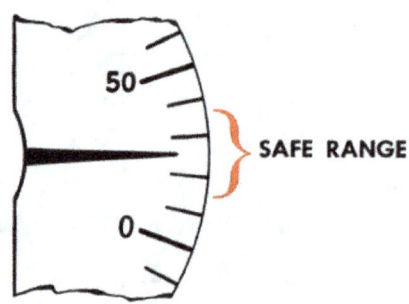

SAFE RULE:
KEEP THE CARBURETOR AIR TEMPERATURE ABOVE 15°C BUT NEVER ABOVE 38°C

retor air temperatures with your intercooler shutters and superchargers.

2. **Intercooler Shutters.** Hot compressed air is coming to your carburetor from the supercharger through the intercoolers. Intercoolers are kept in the "OPEN" position to cool this compressed air. As you climb to higher altitudes it may be necessary to close these shutters to keep the carburetor air temperatures above the icing range. If you do close them, keep a close watch on both carburetor air temperatures and cylinder-head temperatures to be sure that the rise is not beyond limits. **Intercooler shutters should always be used with caution.**

3. **Heater.** Remember that there are crew members all over the airplane who may be getting cold. Ask them if they desire heat. The longer you can keep them warm the more effective they will be with their headwork, their bombs and their guns. Crew comfort is important to crew efficiency.

Decreasing Atmospheric Pressure

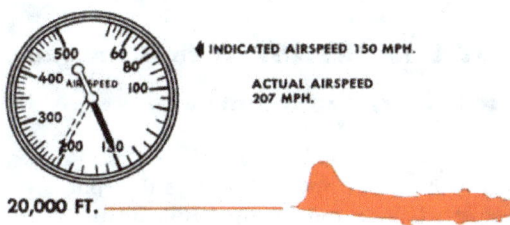

INDICATED AIRSPEED 150 MPH.
ACTUAL AIRSPEED 207 MPH.

20,000 FT.

1. **Airspeed Indicator.** Decreasing atmospheric pressure causes your airspeed indicator to show an airspeed lower than your true airspeed.

2. **Manifold Pressure.** The density and pressure of the outside air is decreasing as altitude increases. At sea level, normal atmospheric pressure on some engines will be sufficient to maintain desired manifold pressure. As altitude increases and full throttles fail to give sufficient manifold pressure, you add boost with the turbo-superchargers.

When climbing at a given throttle setting, rpm, and turbo regulator setting, the manifold pressure will increase slightly with altitude because the atmosphere has less back pressure effect in relation to the constant exhaust pressure. This results in a steady increase in turbo wheel speed.

3. **Rules for Using Turbo-supercharger.**

a. Establish initial manifold pressure with full throttles. Get additional boost from turbo-superchargers.

b. Reduce manifold pressure by first reducing the turbo-supercharger regulators completely and then, if further reduction is necessary, reduce the throttles.

c. At altitude the turbo bucket wheel has a tendency to overspeed. (See pp. 169-173.) The **critical altitude** at maximum power setting of 46″ Hg. and 2500 rpm is 27,000 feet. At 41″ Hg. and 2300 rpm it is 30,000 feet. **Reduce the manifold pressure 1.5″ for every 1000 feet of climb above this critical altitude.** If climbing at less than the maximum manifold pressure, you can raise the critical altitude 1000 feet for each 1.5″ that your manifold pressure is below the maximum. Thus, if the critical ceiling is 27,000 feet at 43″, it will be 29,000 feet at 40″, etc. Then

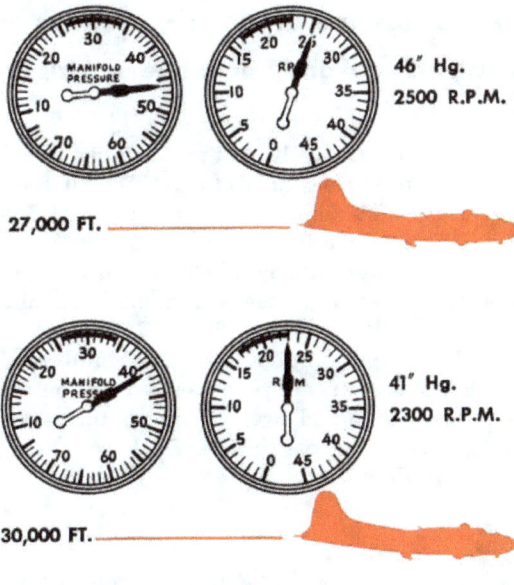

27,000 FT. — 46″ Hg. 2500 R.P.M.

30,000 FT. — 41″ Hg. 2300 R.P.M.

MAXIMUM POWER SETTINGS

continue to decrease manifold pressure 1.5" for each 1000 feet above 29,000 feet.

4. **Booster Pumps on at 10,000 Feet.** As you climb and the atmospheric pressure decreases, there is more and more tendency for suction from your engine-driven fuel pump to cause vapor lock. Booster pumps put 8 lb. additional pressure in the lines to offset this. **Turn the booster pumps on at 10,000 feet** and keep them on until you descend below that altitude.

5. **Crew.** As altitude increases your crew is becoming less efficient. Their ears tend to bother them. Head congestion may cause severe pain and they are getting insufficient oxygen. During day flights, go on oxygen between 7000 and 10,000 feet. At night, have the entire crew use oxygen from the ground up. (See pp. 110-118.)

6. **Carburetor Air Filters.** Avoid use of carburetor air filters above 8000 feet when climbing. Their operation above this altitude will cause rise in carburetor air temperatures, thereby increasing the possibility of detonation. In the B-17F, filters must be closed above 15,000 feet. Failure to observe this precaution may cause detonation and eventual engine failure or sufficient overspeeding of the turbo wheel to cause serious damage. Remember also: use of filters reduces manifold pressure 1" to 2".

The Importance of Smooth Flying

Smooth, steady flying, proper trim, and minimum horsing of the airplane becomes more and more important to maximum performance as altitude increases. Steady, expert flying will cut fuel consumption, eliminate hazards, increase rate of climb, and reduce engine wear.

Remember that the only way you can maintain a constant altitude or climb and smooth, steady flying is with the aid of instruments.

SEQUENCE OF POWER CHANGES

The sequence of power changes for POWER INCREASE is first, mixture controls; second, propellers; third, throttles; last, superchargers.

1. **Mixture Controls:** At the pilot's signal, copilot sets mixture controls to "AUTO-RICH" if necessary. Maximum settings in "AUTO-LEAN" are prescribed for Grade 100 fuel (see Table of Power Setting, p. 86). If power is increased to beyond these maximums, the mixture should be set in "AUTO-RICH" first.

2. **Propellers:** Copilot increases to desired setting. Propellers are set at desired rpm **before** increasing manifold pressure to eliminate the danger of an excessive BMEP (Brake Mean Effective Pressure).

3. **Throttles:** Pilot advances as the rpm is increased. If more power than full throttle is required, advance the superchargers.

4. **Superchargers:** The supercharger controls may be advanced together, but it is advisable to set them one at a time (starting with the dead engine side, if operating with one or more engines dead). Always use full throttle before applying supercharger boost. If throttles are partially closed when turbos are in operation, the resulting back pressure will cause a power loss and possible carburetor damage.

RESTRICTED

The sequence of power changes for POWER REDUCTION is first, superchargers; second, throttles; third, propellers; last, mixture controls.

1. **Superchargers:** Pilot retards supercharger controls slowly in order to prevent cracking of the turbo nozzle box by too rapid cooling. Retard superchargers before throttles to prevent back pressure in the carburetor above the butterfly valve.

2. **Throttles:** Pilot retards throttles. Reason: Manifold pressure must be reduced before propeller rpm in order to keep BMEP on the low side of normal. Although BMEP limits may not be exceeded for a particular case, it is advisable to always use the power sequence so that the pilot will instinctively follow this sequence in emergencies.

3. **Propellers:** Copilot decreases rpm at command of pilot. This must follow throttles to keep sequence in order, as explained above.

4. **Mixture Controls:** Copilot puts mixture controls in "AUTO-LEAN" if new power setting falls within limits.

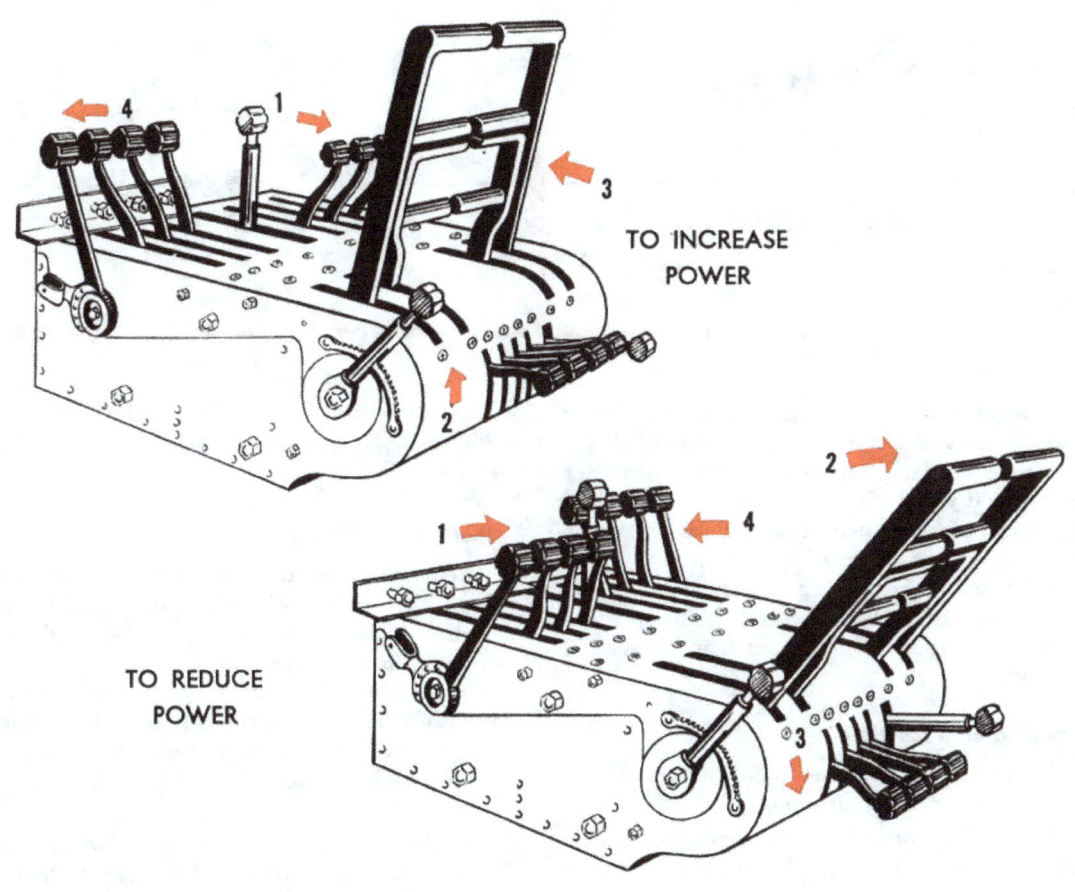

TO INCREASE POWER

TO REDUCE POWER

RESTRICTED 77

LEVELING OFF

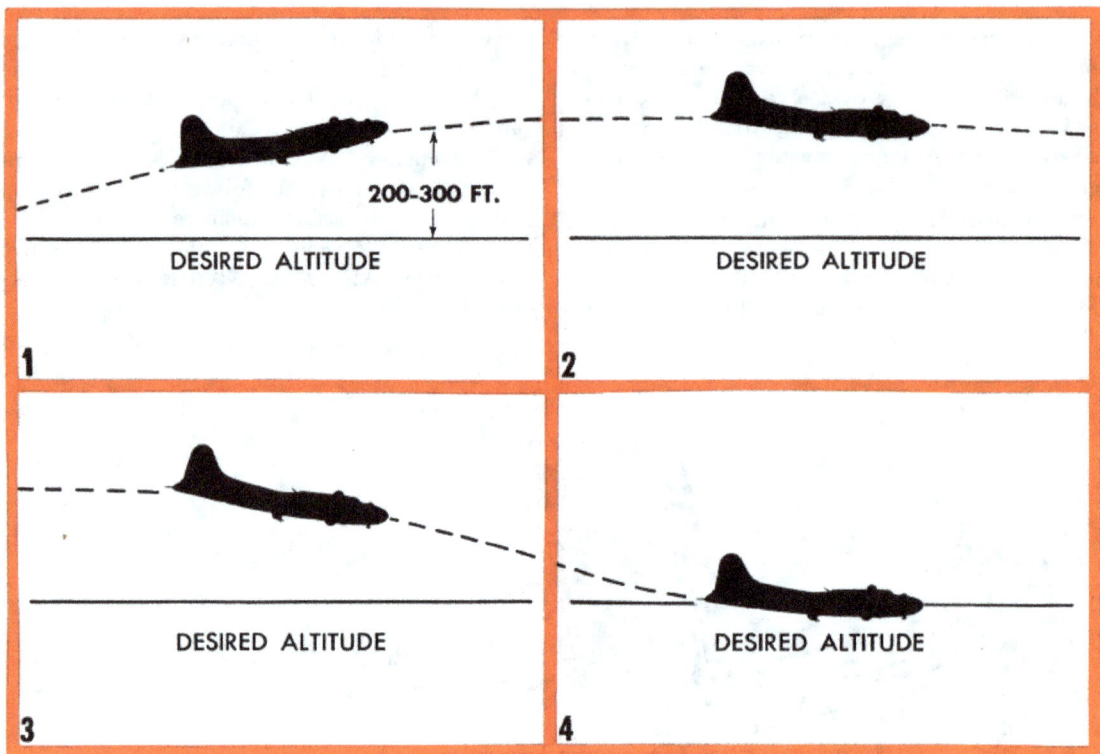

Always level off the cruising **from the top** in both speed and altitude. The purpose of this is to let the airplane build up full momentum for cruising. If you go directly from a climb to level flight and reduce power, the airplane will mush along at a high angle of attack and in a high drag attitude while trying to gain speed. It will fly sluggishly and inefficiently. The heavier your load, the more important it is to level off properly.

Leveling-off Procedure

1. Continue your climb to 200-300 feet above the desired cruising altitude.
2. Level off, drop the nose slightly to get on the step and pick up speed.
3. Reduce power to cruising setting and gradually descend to your cruising altitude.
4. Synchronize propellers and trim the airplane.

Cool Off the Engines

Remember that throughout the climb the engines have been generating heat. Give them a chance to cool down to slightly below desired cruising temperatures before you change to "AUTO-LEAN" mixtures (when using Grade 100 fuel). This allows the cylinders, blower and rear sections to dissipate heat. A well-cooled engine is less likely to detonate than a hot engine.

To aid cooling, don't close the cowl flaps immediately upon completing the climb. Instead close them progressively as airspeed builds up.

TRIMMING

Trimming the B-17 is a routine procedure, but it is tremendously important to the easy and proper operation of the airplane. Brawny, 200-pound pilots have exhausted themselves in one hour's flight merely because they failed to trim properly and frequently enough. Poor trim cuts down airspeed, increases fuel consumption, lowers the speed and ceiling of a climb, and decreases the efficiency of the airplane and the pilot. Formation flying is a nightmare if the airplane is improperly trimmed.

Balance the Power

Make certain that you are using **balanced power**. Propellers should all be synchronized and you should have equal manifold pressure on all engines. **This is important!** Manifold pressure must be equalized exactly to give balanced power.

Elevators

1. Check the flight indicator with the altimeter and rate of climb indicator, and re-set it if necessary for level flight.

2. Hold the airplane level by referring to the flight indicator. Adjust elevator trim to relieve any fore or aft pressure required to hold the airplane level.

Rudders

1. Hold the wings level with the ailerons by reference to the flight indicator and remove all rudder pressure.

2. Watch the directional gyro to see if the airplane is turning. Gradually correct with rudder trim until the directional gyro holds a steady course straight ahead.

Ailerons

1. Level the wings, hold a gyro heading with rudder, and release the wheel.

2. If the flight indicator shows a wing dropping, correct with aileron trim.

Double-check

Finally, check directional gyro, flight indicator, and needle and ball with hands and feet off controls to make sure of proper trim. Once the airplane is properly trimmed, small adjustments will usually keep it there. Trimming should be done automatically, and as quickly as possible. Learn to trim by reference to instruments, and by visual reference to outside objects.

When to Trim

Trim at the first sign of excessive control pressure. You will want to trim for climbs, descent, gear down or up, flaps down or up, when the crew changes positions, as fuel is used up, when bombs are dropped, in case of engine failure, when cowl flaps are changed, etc.

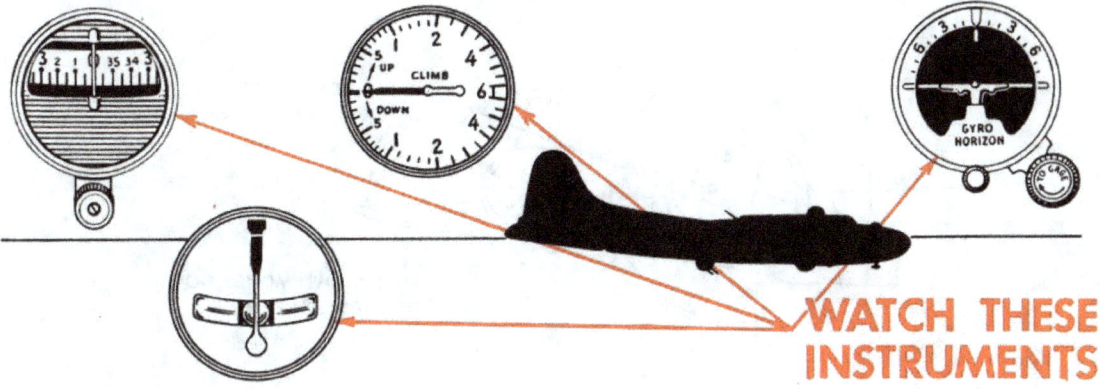

WATCH THESE INSTRUMENTS

HOW TO SYNCHRONIZE PROPELLERS

The copilot brings propellers to desired tachometer setting with the propeller governor controls. Although rpm readings may be identical for all four engines, propellers may not be perfectly synchronized because of slight variations in tachometers.

Procedure for Synchronizing

1. To synchronize No. 1 and No. 2 propellers, leave No. 2 rpm unchanged. Have navigator or some crew member in the nose look at the propellers and report the direction of the **rotating shadow** (where the propellers appear to overlap). If the shadow is moving in the direction **opposite** to No. 1 propeller rotation, (up) that propeller is too slow and rpm should be increased. If the shadow is moving in the **same** direction as No. 1 propeller rotation, No. 1 propeller is too fast and should be decreased.

2. To synchronize No. 3 and No. 4, leave No. 3 rpm unchanged. If the shadow is moving in the direction **opposite** to No. 4 propeller rotation, the No. 4 propeller is too slow and rpm should be increased. If the shadow is moving in the same direction as No. 4 propeller rotation, it is too fast and should be decreased.

Check the Shadow

Remember that as seen from the pilot's seat all four propellers rotate to the right. Thus No. 1 turns toward you and No. 4 away from you. If the shadow rotates with the propeller, the propeller is too fast. If it rotates backward (against the propeller rotation) the propeller is too slow.

1. Make small adjustments with propeller controls. When propellers are synchronized, shadows will disappear.

2. If shadows have disappeared and the engines still sound unsynchronized, (a distinct pulsation or engine beat) then the two propellers on one side are not synchronized with the two on the other side.

3. Synchronize left propellers with right propellers. Check the tachometers to see if either pair is indicating less than the desired rpm. If so, make small adjustments with the two propeller controls until you eliminate the beat and get a steady drone. If the beat gets worse, decrease rpm instead of increasing.

The difference in tachometer needle travel will indicate which governors are slow. With practice you will be able to lead with the controls for slow-acting governors and bring all four propellers to desired rpm simultaneously.

Synchronizing at Night

Use landing lights or flashlight to determine the rotation of shadows. With practice, you can complete the entire operation by sound.

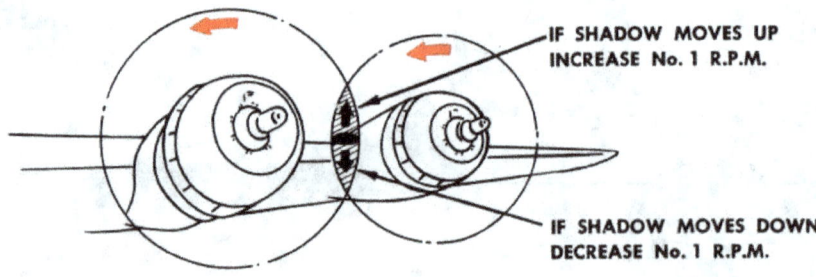

Cruising

As soon as you have leveled off, synchronized propellers, trimmed the airplane, and let the engines cool down, check all instruments before going into "AUTO-LEAN." **(Auto-lean is used only when operating on Grade 100 fuel.)**

Normal Pressures and Temperatures for Automatic Lean

1. Cylinder-head temperature: 218°C maximum; desired 205°C or below.
2. Oil temperatures: 75°C desired; 88°C maximum.
3. Oil pressures: 70 to 80 lb. sq. in.
4. Fuel pressures: 14 to 16 lb. sq. in.
5. Carburetor air temperature: from 15°C to 38°C.

Automatic Lean

If instrument readings are satisfactory, copilot (at the pilot's direction) moves the mixture controls one at a time to "AUTO-LEAN." Pilot and copilot note the effect of this on temperatures and pressures.

Carburetor air temperature should be kept below 38°C, as excessive heat may cause detonation. If an engine gets hot in "AUTO-LEAN" (a less cooling mixture) go to "AUTO-RICH" long enough to cool it down. If it stays hot in

"AUTO-LEAN," the automatic feature may not be operating properly, and you may have to use "AUTO-RICH" for that engine.

Booster Pumps

Remember to keep booster pumps on when cruising above 10,000 feet.

Superchargers

Low altitude: If cruising at a low altitude you may have sufficient manifold pressure with superchargers completely off. However, under icing conditions and extremely cold air temperatures, it is important to keep superchargers engaged and operating to prevent induction system icing and to insure warm oil to the supercharger regulators.

1. Engage superchargers and increase desired manifold pressure 1½".
2. Reduce throttles to re-establish desired manifold pressure.

Above 20,000 feet: Superchargers won't function properly when engines are operating at less than 1800 rpm above 20,000 feet, because in thinner air there is insufficient exhaust gas to operate the turbo wheel at the necessary speed. Don't suspect turbo regulator trouble until you have checked rpm.

Cowl Flaps

Regulate cylinder-head temperatures with cowl flaps. The closed position reduces drag and increases speed.

Directional Gyro

Check and correct for precessing at least every 15 minutes, or as often as necessary.

Note: Although pilot and copilot will be checking instruments regularly, it is a good idea to call for a complete check and report by the copilot at regular intervals.

Flying the Airplane

Take pride in your ability to fly the airplane as perfectly as possible. You can't expect your copilot or your crew to develop keen interest in the technique of their jobs unless you set an outstanding example.

Heading

Hold your heading. If you are going to change heading, or dive or climb, warn your navigator in advance exactly what to expect. Know where you are, but let the navigator navigate. Require position reports every 30 minutes.

Altitude

Hold your altitude. Don't be satisfied with 200 feet higher or lower.

Airspeed

As time passes and your fuel load lightens, your airplane will tend to gain airspeed. Maintain your recommended IAS (i.e., 150-155 mph for long-range cruising) by reducing rpm every 1 to 3 hours. This is always a good rule for efficient cruising.

Automatic Pilot

See section on use and operation of automatic pilot, pp. 183-190.

Flight Performance Record

It is the copilot's duty, with the assistance of the engineer, to keep a flight performance record of every mission. (See suggested form.) Preferably entries should be made every 30 minutes. Properly kept, this form will:

1. Warn you of excessive gas consumption.
2. Give a running report of the performance of engines.
3. Provide a check on how efficiently you are flying the airplane.

Engineer's Hourly Visual Check

Require the engineer to make a visual check once an hour of instruments, engines, nacelles, fuel cell areas. Above 15,000 feet this check can be postponed at the direction of the pilot for purposes of crew safety.

Oxygen

When on oxygen require the copilot to check crew stations once every 15 minutes by interphone to ascertain that crew members are all right and have an adequate supply of oxygen.

RESTRICTED

FLIGHT PERFORMANCE RECORD

Airplane No.　　　　　Date　　　　　Pilot

Time of T.O.　　　　　To　　　　　Copilot

Time	I.A.S.	Alt.	Free Air Temp.	Fuel in Tanks								Fuel Consumed in Last Period		RPM			
				1	2	3	4	LBB	RBB	L. Aux.	R. Aux.	Gals.	Gals. Per Hr.	1	2	3	4

FLIGHT PERFORMANCE RECORD—Continued

Wt. at T.O.　　　　　Total Oil at T.O.　　　　　Mission

C.G. at T.O.　　　　　Total Fuel at T.O.　　　　　From:

Manifold Pressure				Mixture Control				Cyl. Head Temp.				Oil Pressure				Oil Temperature				Wt.	Pos. of C.G.
1	2	3	4	1	2	3	4	1	2	3	4	1	2	3	4	1	2	3	4		

RESTRICTED

Long-Range Cruising

For normal long-range cruising (with all engines operating, and no external loads):

1. Below 20,000 feet set rpm to maintain an IAS of 150 mph, manifold pressures of 29" (+ or — 1") Hg., 1400-2000 rpm as required.

2. Above 20,000 feet use 29" (+ or — 1") Hg. and an rpm necessary to maintain an IAS of 140 mph.

3. If long-range cruising speed cannot be maintained up to 2000 rpm, use higher rpm with correspondingly higher recommended manifold pressures.

4. With Grade 100 fuel, at or below 2100 rpm use "AUTO-LEAN" mixture.

5. Close cowl flaps or adjust for proper cylinder-head temperature (205°C or below).

6. Hold power setting and let airspeed increase up to 155 mph as fuel is used. Re-set rpm every 3 hours to maintain desired cruising speed.

Reason why airspeed is maintained at 140 mph: power necessary to maintain 150 mph increases with altitude to a point where "AUTO-RICH" mixture becomes necessary (when using Grade 100 fuel only) unless airspeed is reduced, thereby using more fuel.

For long-range cruising (a) **with one engine dead**, (b) **with 2 engines dead**, (c) **with all operating, carrying extra bomb loads**:

1. Use the same manifold pressure as stated above.

2. Fly at 145 mph IAS below 20,000 feet.

3. Fly at 130 mph above 20,000 feet.

The reduced airspeed is necessary under these conditions because of increased power requirements, and for the same reason as at high altitudes: higher airspeed would require rich mixtures and cause engine inefficiencies.

Maximum Endurance

To remain aloft for the greatest possible length of time with a given amount of fuel (and where distance flown is no consideration), you will have to employ a technique considerably different from that used for long-range cruising. This technique is sometimes called **hovering**.

Calculating the approximate variations in power required for level flight in a medium gross weight B-17: **minimum** power is required at airspeeds around 110 mph; only slightly more power is required at 120 mph; whereas substantially more power is required at 130 mph.

Therefore, the best and safest hovering (maximum endurance flight) can be done at 120 mph, since there is reserve speed and fuel consumption is only slightly more than at 110 mph.

1. Using no flaps, set manifold pressures at 29" Hg., rpm as required down to 1400 rpm, and keep an airspeed of 120 mph.

2. If lower power than 29" Hg. and 1400 rpm is needed to maintain 120 mph, reduce manifold pressures to 26" Hg.

3. Do not feather any engines.

4. As in cruising, the lower altitudes will yield the best performance. Reasonable altitude (several thousand feet above the ground), obviously, must be maintained.

Reference: Flight Operation Instruction Charts, and Composite Cruising Control Chart, AN 01-20-EF-1 and AN 01-20-EG-1.

POWER SETTINGS
FOR GRADE 100 AND GRADE 91 FUEL

The 2 accompanying charts present a clear picture of the engine operating limits for Wright R-1820-97 engines using Grade 100 and Grade 91 fuel.

The charts are divided into 3 **regions of operation**. The "Desirable Region of Operation" is in the center, and is based on the allowable limits within which a given combination of manifold pressure and engine rpm will produce economical fuel consumption and avoid pre-ignition and detonation. To the left of the "Desirable Region" lies the "Prohibited Region" where excessive engines pressures cause pre-ignition and detonation. To the right of the "Desirable Region" lies the "Region of Excessive Consumption."

Desirable Region of Operation

The meaning of this designation is obvious: If the pilot chooses to operate outside the region indicated, he can expect the consequences—detonation or excessive fuel consumption.

The black points between the lines indicate the recommended power settings. Notice that 1" of manifold pressure more or less is considered allowable.

Note
ON THE USE OF GRADE 91 FUEL

The principal concern of the pilot operating on a different grade of fuel than that for which the engine was designed should be the possibility of detonation.

The power settings in the "Desirable Region of Operation" on the accompanying chart are recommended for the best all-round engine performance when operating on Grade 91 fuel. Exceeding these manifold pressures at a given rpm **will result in detonation and undue stress on the engine;** operating at a lower manifold pressure for a given rpm **will result in excessive fuel consumption.**

A dead sparkplug can cause detonation, which will develop into pre-ignition on the side of the piston where the dead sparkplug is installed.

Bear in mind the relationship of manifold pressure to engine revolutions. The settings recommended for a given power output are **minimum rpm** and **maximum pressure**. While an increase in rpm and a reduction in manifold pressure would result in a condition more favorable to long engine life, it would also result in excessive fuel consumption.

Detonation and Pre-ignition

Detonation may be described as the condition in which the fuel charge in the cylinder fires spontaneously and too rapidly instead of progressively burning over a longer period of time.

Pre-ignition is one of the results of detonation. Local hot spots within the combustion chamber (excessive carbon or other deposits) reach such high temperatures that they cause ignition of the fuel-air mixture before it can be ignited normally by the ignition spark. Pre-ignition is even more severe than detonation itself in its effect on the engine. The engine will not continue to operate for more than a short time when pre-ignition is present.

Results of detonation or pre-ignition are: (1) reduction in power output, (2) possible engine failure, (3) actual damage to the engine parts.

Detonation may be the result of a variety of causes (see T.O. No. 02-1-7) and may occur

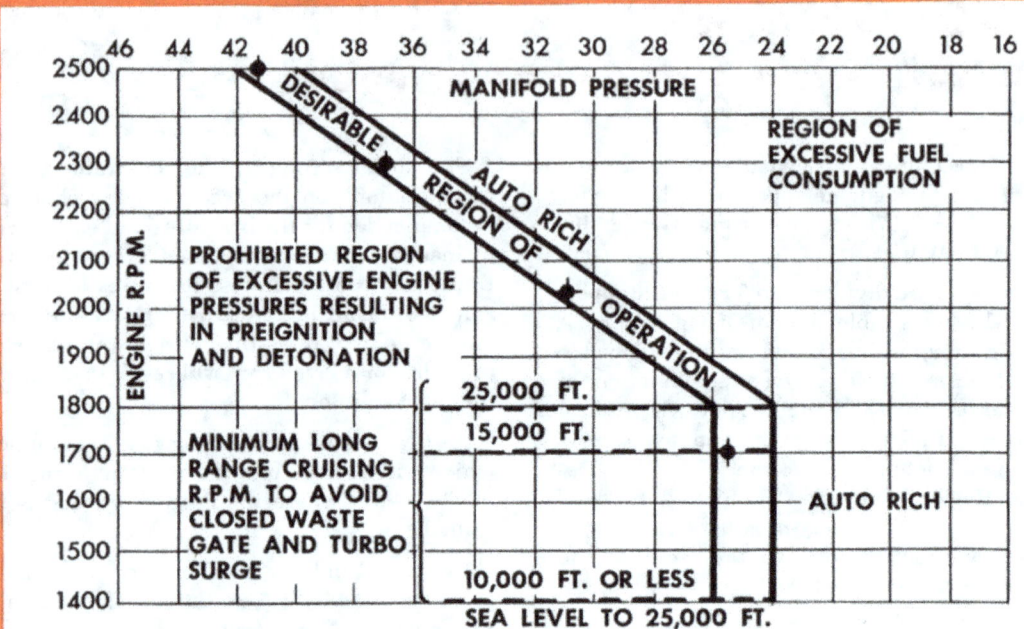

POWER SETTINGS FOR GRADE 91 FUEL

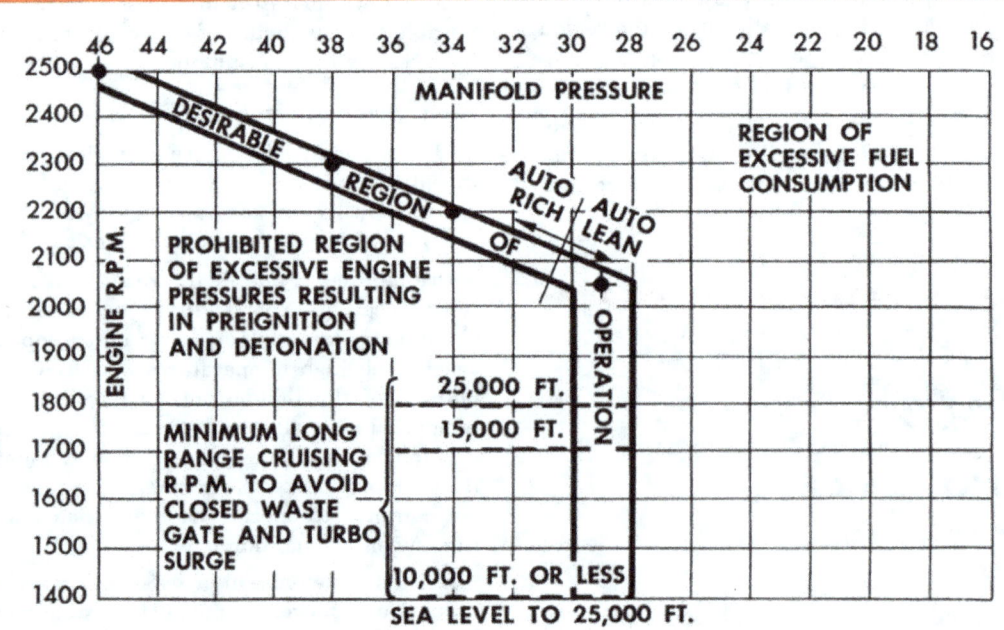

POWER SETTINGS FOR GRADE 100 FUEL

at any of a great number of manifold pressure and power settings depending upon the octane rating of the fuel, the original temperatures of gasoline, carburetor air, cylinder heads, etc. Therefore, no definite lines can be drawn on the charts to show exactly where detonation will occur.

The only safe operation procedure is to stay within the "Desirable Region of Operation" or, conversely, to stay out of the regions of excessive engine pressures where detonation may occur.

FOR GRADE 91 FUEL: OPERATING LIMITS OF THE B-17

Condition	RPM	Manifold Pressure	Mixture	Estimated Fuel Consumption*
Takeoff	2500	41" Hg.	AUTO-RICH	550 gal/hr.
Maximum Climb	2300	37" Hg.	AUTO-RICH	480 gal/hr.
Desired Climb	2300	35" Hg.	AUTO-RICH	435 gal/hr.
Maximum Cruise	2020	31" Hg.	AUTO-RICH	280 gal/hr.
Normal Cruise	2000	28" Hg.	AUTO-RICH	200 gal/hr.
Long Range	150 mph IAS	28" Hg.	AUTO-RICH	200-140 gal/hr.

FOR GRADE 100 FUEL: OPERATING LIMITS OF THE B-17

Condition	RPM	Manifold Pressure	Mixture	Estimated Fuel Consumption*
Takeoff	2500	46" Hg.	AUTO-RICH	560 gal/hr.
Maximum Climb	2300	38" Hg.	AUTO-RICH	440 gal/hr.
Desired Climb	2300	35" Hg.	AUTO-RICH	400 gal/hr.
Max. Cruise	2200	34" Hg.	AUTO-RICH	350 gal/hr.
Max. Cruise	2100	31" Hg.	AUTO-LEAN	250 gal/hr.
Normal Cruise	2000	29" Hg.	AUTO-LEAN	188 gal/hr.
Long Range	150 mph IAS	29" Hg.	AUTO-LEAN	188-122 gal/hr.

*Operation on all 4 engines

FLIGHT CHARACTERISTICS

The B-17F possesses many outstanding flight characteristics, chief among which are: (1) directional stability; (2) strong aileron effect in turns; (3) ability to go around without change in elevator trim; (4) exceptionally satisfactory stalling characteristics; and (5) extremely effective elevator control in takeoff and landing.

Trim Tabs

The airplane will go around without changes in elevator trim tab settings. However, trim must be changed with adjustment of cowl flaps and power settings, for these reasons:

1. Increased power on the inboard engines causes the airplane to become slightly tail-heavy. (Power change on the outboard engines has no appreciable effect on trim.)

2. Closing the cowl flaps on the inboard engines also causes tail-heaviness. (The effect of cowl flaps on the outboards is negligible.)

With the airplane properly trimmed for a power-off, flaps-down landing, you can take off and go around again by applying power and putting the flap switch "UP" with no change in trim. The flaps will retract at a satisfactorily slow rate.

Turns

Because of the inherent directional stability of the B-17, dropping one wing will produce a noticeable turning effect. Very little rudder and aileron will enable you to roll in and out of turns easily. Carefully avoid uncoordinated use of aileron.

In shallow turns the load factors are negligible. But in steeper turns proportionately more back pressure is required, thereby increasing the load factor.

In banks from 10° to 70° the load factor increases from 1.5 to 3.0. Obviously, steep turns of a heavily loaded airplane may place sufficient stress on the wings to cause structural failure.

If the airplane tends to slip out of turns, recover smoothly without attempting to hold bank. Decrease the bank. Use proper coordination of rudder and aileron.

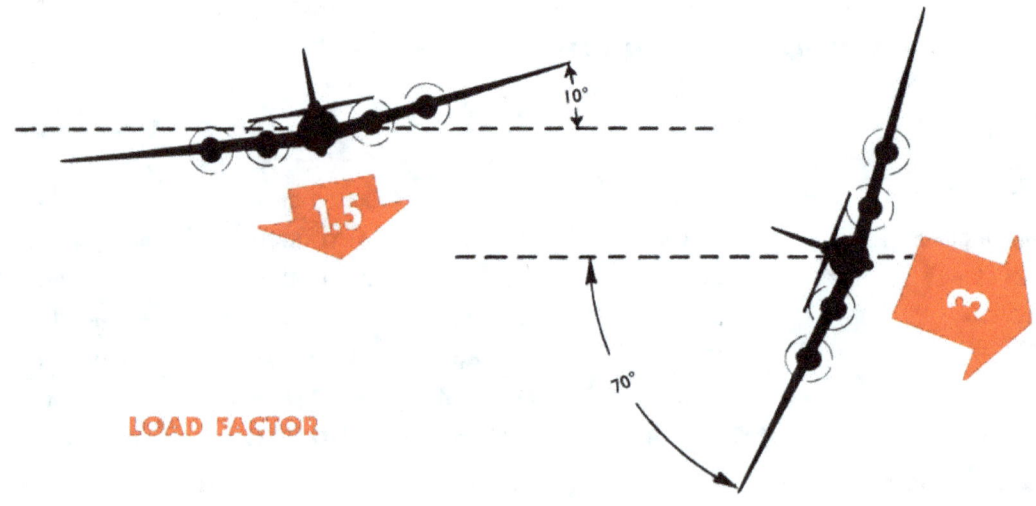

LOAD FACTOR

Rough Air Operation

In rough air, use both rudder and ailerons without worrying about excessive loads. Both aileron and rudder forces vary with changes in airspeed in such manner that it is almost im-

possible to damage the system without deliberately trying to do so. Necessary control pressures are small enough, and the responses large enough, to maintain ample control of the airplane.

However, in the case of the elevators, exercise great care, both in rough air and in recovery from dives, to assure smooth operation. In thunder storms, squalls, and in or near turbulent cumulus clouds, it is possible to develop excessive load factors by means of the elevators unless they are used properly. This does not mean that there is any greater tendency to exceed allowable load factors in the B-17 than in other heavy bombardment or transport airplanes. It means that **the larger the airplane, the greater the time and distance required to complete any maneuver.** In operation, you **must** allow more distance and time in proportion to the size of the airplane.

Generally, in rough air, hold constant airspeed by means of the elevator, but do it smoothly. Remember that recovery to the desired airspeed may take time.

Avoid hurried recovery from dives, climbs or changes in airspeed. Never dive the airplane through a cloud layer or through rough air at maximum diving speed. Don't attempt high-speed flight in rough air.

Stalls

The stall characteristics of the B-17 are highly satisfactory. The tendency to roll—commonly caused by lack of symmetry in the stalling of either wing—is minimized by the large vertical tail. Under all conditions a stall warning at several mph above stalling speed is indicated by buffeting of the elevators.

If airspeed is reduced rapidly near the stall, the speed at which the stall will occur will be lower than when the stall is approached gradually. The stall will also be more violent because the wing's angle of attack will be considerably above the stalling attitude.

The stalling speed of the B-17F, like that of any other airplane, depends upon: (a) the gross weight, (b) the load factor (number of Gs), (c) the wing flap setting, (d) the power, (e) de-icer operation and ice formation.

The effect of gross weight upon stalling speed is obvious: the heavier the load, the higher the stalling speed.

The effect of the load factor is simply to increase the effective gross weight in proportion to the load factor.

The greater the flap angle the lower the stalling speed. The greater the power, the lower the stalling speed. Full flaps reduce the stalling speed about 15 mph for gross weights of 40,000 to 45,000 lb., and a load factor of 1.0; but full military power for the same loading conditions may reduce the stalling speed another 15 mph.

Any yawing, accidental or otherwise, will increase the stalling speed and any tendency to roll at the stall. This is obvious, since the normal procedure in deliberately making a spin is to yaw the airplane as it stalls. For example, if the left wing drops at the stall and you apply right aileron to raise the left wing, the ailerons will have a tendency to overbalance and **reverse** effectiveness, because of the drag induced by the aileron. The result will be increased dropping of the left wing. The aileron procedure in recovering from a stall, therefore, is to **hold ailerons neutral and refrain from their use until coming out of the dive in the final phase of recovery.**

RECOVER FROM STALL SMOOTHLY

Stall Recovery

For the B-17F the procedure for recovering from a stall is normal.

1. Regain airspeed for normal flight by smooth operation of the elevators. This may require a dive up to 30°.

2. While regaining airspeed, use rudder to maintain laterally level flight. After airspeed is regained, use ailerons also for lateral control—**but not until airspeed is regained.**

The important thing is to **recover from the dive smoothly.** Penalty for failure to make a smooth recovery may be a secondary stall or structural damage to the airplane, both because of excessive load factors. Rough or abrupt use of elevators to regain normal flying speed may cause the dive to become excessively steep.

The additional airspeed necessary to regain normal flight need not be more than 20 mph. This means that excessive diving to regain airspeed is absolutely unnecessary.

Remember these additional facts about stalls:

1. Stalls with wheels down will increase the stalling speed about 5 mph.

2. Stalls with wheels and flaps down will decrease the stalling speed about 10 mph.

3. Stalls with de-icer boots operating will increase the stalling speed 10-15 mph. In recovering from stalls with de-icer boots operating, regain slightly more than the usual 20 mph needed for recovery. Such stalls are apt to be more abrupt, with a greater tendency to roll.

Spins

Accidental spinning of the B-17 is extremely unlikely. The directional stability and damping are great, and it is probable that even a deliberate spin would be difficult. However, remember that **the airplane was not designed for spinning, and deliberate spins are forbidden.**

Dives

The maximum permissible diving speed in the B-17F (flaps and wheels up) with modified elevators is 270 mph IAS; without elevator modifications, the maximum diving speed is 220 mph.

The structural factors limiting the diving speed of the B-17F are the engine ring cowl strength, the wing leading-edge de-icer boot strength, the cockpit windshield and canopy strength, and the critical flutter speed. The engine ring cowl has been designed to withstand 420 mph. The windshield and cockpit canopy have ample margin at 305 mph. The wing leading-edge de-icer boots begin to raise slightly from the wing at 305 mph, and any additional speed would be likely to lift the upper part of the boot above the wing surface, possibly causing structural failure. The mass balance of the control surface is so essentially complete both statically and dynamically that, basically, the critical flutter speed depends entirely on the wing-bending torsion critical speed, which is approximately 375 mph.

Therefore, it is obvious that simply diving the airplane (with modified elevators) to 270 mph involves no danger whatsoever. The only danger that must be considered is in recovery. Recovery must be smooth and gradual. Normally, a load factor of 2 will not be exceeded. At the gross weight of 50,000 lb., the initial-yield point factor is slightly less than 3, making the ultimate load factor slightly over 4. Obviously, at that gross weight the load factor 3 should never be reached; the load factor 2 normally will not be exceeded.

Heavy Loads

The B-17 is stable longitudinally with heavy loads as long as the center of gravity is forward of 32% of the Mean Aerodynamic Chord (87 inches aft of the leading edge of the center section).

For all normal loading the CG must be kept forward of 32% of the MAC. If an excessive load is placed in the rear, the airplane will have neutral or negative stability. It is possible to trim the airplane with an unstable load, but it will be difficult to fly, especially on instruments. It is also much easier to stall inadvertently when flying an unstable airplane on instruments.

Loading for the forward CG positions is preferred because, in addition to being easier to fly, it gives a smooth increase in elevator forces required to pull out of dives, and eliminates the necessity of using excessive elevator trim to hold the tail up.

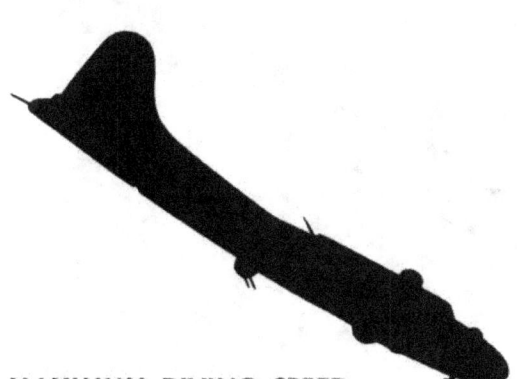

MAXIMUM DIVING SPEED WITHOUT ELEVATOR MODIFICATIONS IS 220 MPH.

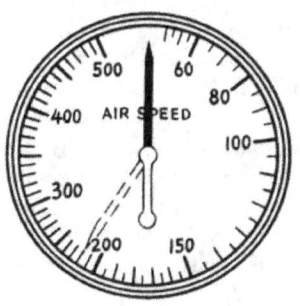

RESTRICTED
LANDINGS

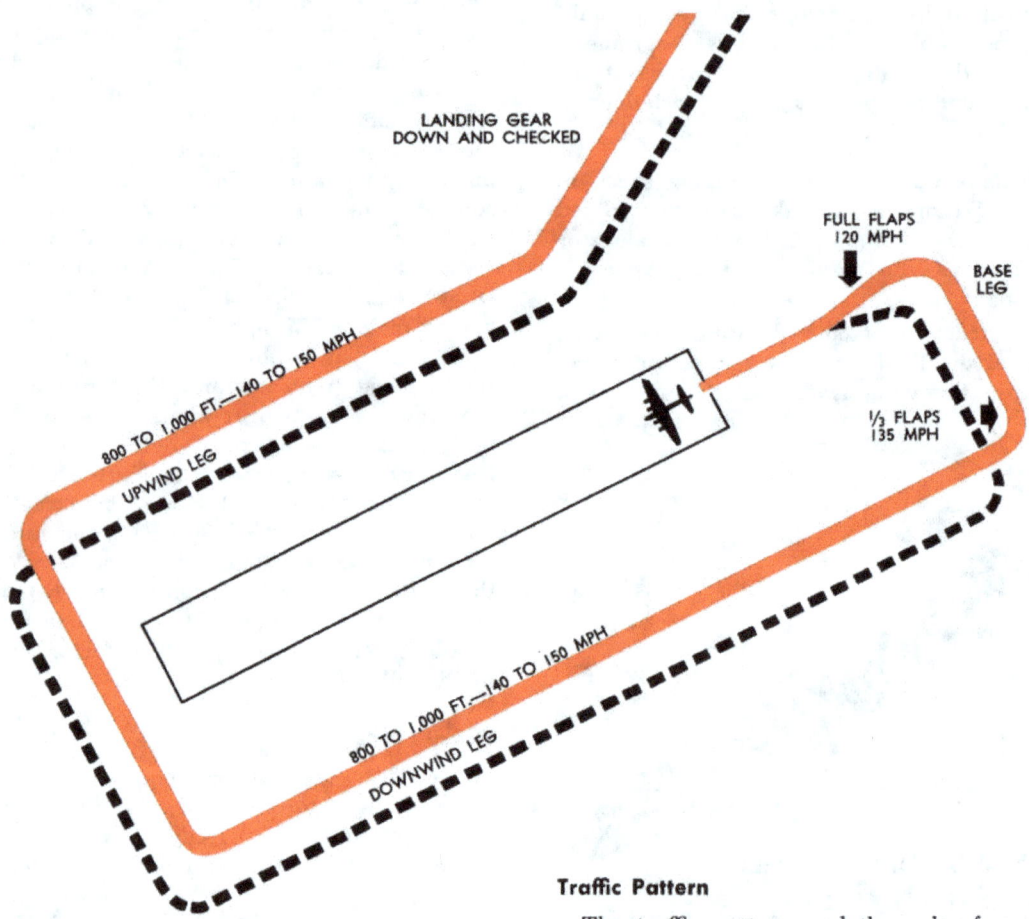

The before-landing check (see Pilot's Checklist) is used when returning from a mission that takes the airplane away from the home field, i.e., for other than traffic pattern work. Complete this check before entering the traffic pattern, so that thereafter you will be able to devote your undivided attention to traffic and landing. (For traffic pattern work a subsequent landing check is provided. See pp. 55-56.)

Traffic Pattern

The traffic pattern and the rules for entering and flying it are prescribed by local field regulations. At the majority of B-17 stations within the continental U.S., the pattern is rectangular in shape. The pattern altitude may vary, but generally it is between 800 and 1000 feet above the ground.

For traffic and safe spacing purposes, fly the pattern at 140-150 mph IAS and 2100 rpm, with manifold pressures sufficient to hold the desired airspeed, but not in excess of 31" Hg. (If more power is needed, increase rpm and manifold pressure together.) When 1/3 flaps are lowered when turning on base leg, maintain an airspeed of 135 mph.

BEFORE-LANDING CHECK

Radio Call, Altimeter Setting

Radio call to the tower is made by pilot or copilot (see **Pilot's Information File**). Obtain altimeter setting for the field and landing instructions. Repeat the altimeter setting to the tower to insure correctness. (Final radio call will be made while in traffic.)

Crew Positions

Have the engineer check the crew to see that all members are in proper positions for landing.

The radio operator will check the trailing antenna and see that it is retracted.

Gunners will check their guns and make sure they are in proper position for landing.

Automatic Pilot

See that the automatic pilot is "OFF." All switches must be turned "OFF" to eliminate any possibility of accidental engagement.

Booster Pumps

Check the booster pumps "ON."

Intercoolers

Be sure the intercoolers are in the "OFF" or "COLD" position for landing. Intercoolers "ON" will cause detonation and loss of power if emergency power is needed on the landing.

When freezing precipitation is present during the approach glide to the runway, and there is danger of carburetor icing, turn the intercoolers "ON," but be sure to notify the copilot and all persons on the flight deck. This will serve as a reminder to all that, in any emergency, the intercoolers must be turned "OFF" immediately.

Carburetor Filters

Place the carburetor filters in the "ON" (or "OPEN") position for landing. With filters off, or closed, a rise in available manifold pressure takes place. If left off or closed for landing, dangerous manifold pressures will develop should emergency power or full throttle be used.

Wing De-icer Boots

Check the wing de-icer boots: controls should be in the "OFF" position except when testing or actually in use.

Make a visual check to be sure the de-icer boots are deflated before the final approach. Remember that action of the wing de-icer boots disturbs the flow of air over the lifting surfaces and materially increases the stalling speed.

Check propeller anti-icers: "OFF." The rheostats of the propeller anti-icers usually are set at a predetermined rate of flow. Their adjustment should not be changed.

Landing Gear

Instruct the copilot to put the landing gear switch in the "DOWN" position. Make a visual check from the left-hand window, and report aloud: "Down left." The copilot will make a similar check on his side and will report, "Down right." From the rear of the airplane, the engineer will check the tailwheel and report: "Tailwheel down." At the same time, the engineer will visually check the condition of the tailwheel assembly (no worn threads or gear, etc.), and see that the trailing antenna is retracted. Engineer will check the ball turret.

Check Landing Gear Warning Lights

Copilot returns switch to neutral position and checks warning light: green light on.

Hydraulic Pressure

With landing gear down, check the hydraulic pressure gages: normal pressure is 800 lb.

Service the accumulators, if necessary.

Be sure the cowl flap controls are in the "LOCKED" or neutral position to prevent any loss of oil supply through leaks in the actuating mechanism.

If in doubt about hydraulic pressure, instruct the copilot to stand by on the hand pump, awaiting your signal.

Increase RPM

In the traffic pattern, signal the copilot to increase rpm to 2100.

Turbos

Decrease manifold pressure to about 23", then signal the copilot to place the turbo controls full "ON," readjust manifold pressure to desired value. Be extremely careful that allowable manifold pressures are not exceeded with the turbos in the full "ON" position. This is important in case an emergency takeoff or go-around is necessary after an attempt landing. Normally, full takeoff manifold pressure will not be needed in such an emergency, since the airplane still will be at or near flying speed and no original inertia has to be overcome.

Flaps

Lower ⅓ flaps when turning on base leg, after airspeed has been reduced below 147 mph.

FINAL APPROACH

Flaps

For normal landings, place the wing flaps in the full down position on the final approach. However, in heavy winds or heavy crosswinds partial flaps produce better results.

In the event of an emergency takeoff or go-around after an attempted landing, do not retract flaps until full power has been applied.

High RPM

While fully retarding throttle, signal the copilot to move propeller controls to full "HIGH RPM."

Power-off Approach

The power-off approach can be executed successfully in normal empty-weight B-17's, and is taught in transition schools.

The important factors in making a successful power-off approach are: (1) setting the proper base leg—not more than 3 miles out; (2)

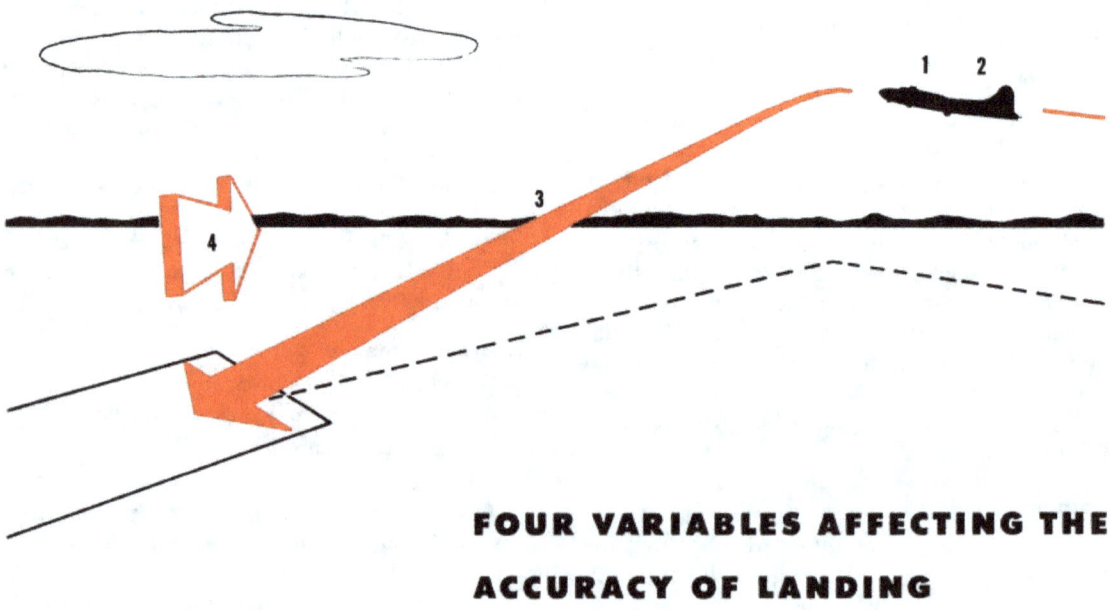

FOUR VARIABLES AFFECTING THE ACCURACY OF LANDING

maintaining constant altitude on the base leg; (3) maintaining constant airspeed and angle of glide; and (4) the wind.

These are the 4 variables. The first 3 are under your control; the fourth—the wind—can be taken care of by proper application of the first 3 factors.

Usually, a good approach means a good landing. The best approach can be made by setting the base leg approximately 2 miles from the field, never more than 3 miles.

Maintain altitude throughout the turn on the approach.

The third and most important consideration in the successful approach and landing is to maintain a constant glide. Roll out of the turn on the approach, lower flaps, maintain altitude, and reduce power at the proper point. Smoothly blend power reduction to the change to gliding attitude.

A good or bad landing of a 4-engine airplane usually is determined by the time it has descended to 300 feet. By that time the pilot should have established constant glide, constant airspeed, constant rate of descent, and made an accurate judgment of distance. If he has accomplished these things, the landing is in the bag.

Proper altitude for breaking a power-off glide is approximately 150 feet with a medium load. The flatter the glide, the lower the glide may be broken.

Level off for landing smoothly and gradually. In the B-17 an abrupt change of attitude from the vertical to the horizontal plane will increase the wing loading, thereby increasing the stalling speed. There is no danger of this if you level off smoothly and gradually.

Power Approach

The same 4 variables—setting the base leg, maintaining altitude on the base leg, holding constant airspeed and angle of glide, and reckoning with the wind, govern the success or failure of the power approach.

The power approach does **not** mean flying the airplane in at excessive speed and skimming over half the runway's length. Nor does it mean bringing the airplane in at such a low speed that it is virtually hanging on the prop to stall in as soon as throttles are cut.

The power approach is a **controlled glide** in which power is used to obtain accuracy in landing on a selected spot, and greater control of the airplane.

Put down flaps and reduce power on the approach. Continue to reduce power **gradually** until the desired airspeed and rate of descent have been established. (Approximately 15" Hg. on the B-17E, and approximately 20" Hg. on the B-17F and B-17G). Hold a desired manifold pressure until you are ready to close throttles when nearing the runway. This eliminates any need for jockeying throttles back and forth, and makes for a smooth, precise landing.

Normally, the gliding speed should be maintained at 120 mph; but this will vary with the gross weight and CG location, rigging, angle of descent, wind conditions, and pilot technique. Correct glide usually results from bringing these factors into harmonious relationship. Proper gliding speed is approximately 20% above stalling speed for a B-17 with a medium load.

Strong Winds

When landing in strong winds, the use of full flaps often is inadvisable. Use your discretion as to the amount of flaps to use. However, never use less than ½ flaps.

Crosswinds

When turning on the approach in a crosswind, be careful to prevent the wind from forcing you off your approach to a degree where it is impossible to align with the runway.

There are 3 possible ways of making a crosswind approach and landing: **(1) holding the** airplane straight toward the runway, dropping one wing into the wind with just enough top rudder to counteract drift; (2) heading the airplane into the wind (crabbing) just enough to keep a straight ground path; and (3) a combination of the first two methods.

The last combination of methods is preferred, because it eliminates the possibility of dropping the wing too low, or of crabbing too much. It also prevents crossing controls and

CROSSWIND LANDING TECHNIQUES

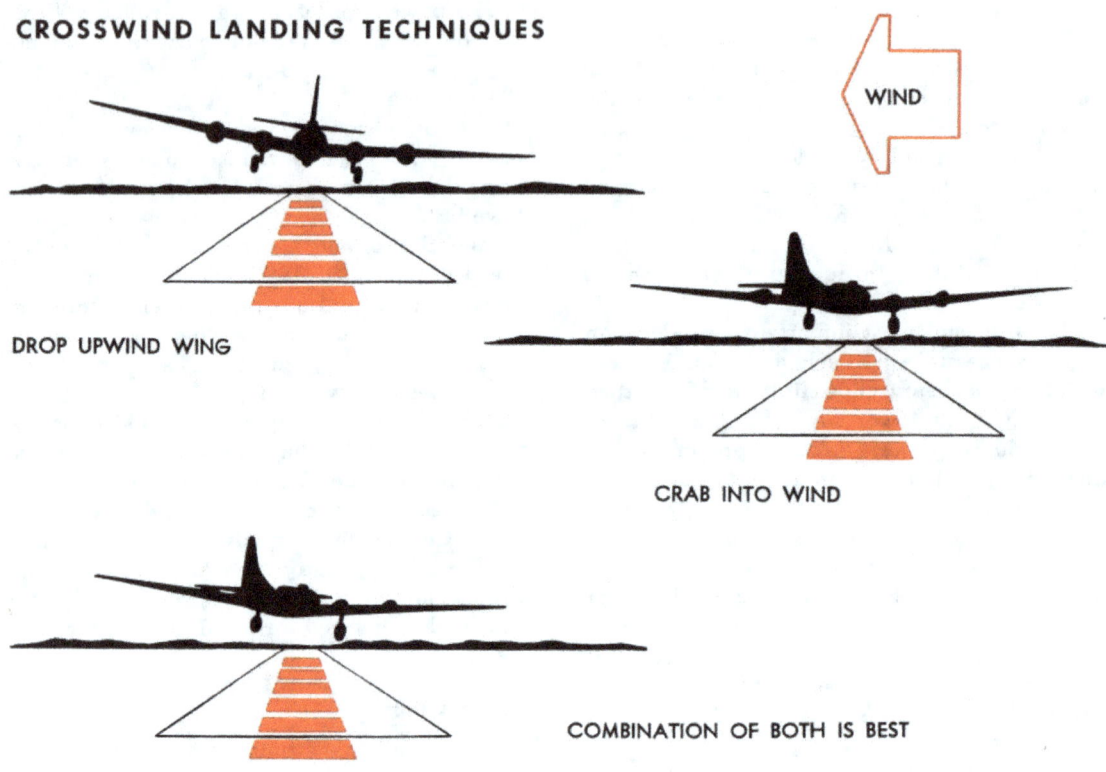

DROP UPWIND WING

CRAB INTO WIND

COMBINATION OF BOTH IS BEST

---

decreases the amount of correction needed to straighten out and level off during the round-out.

If the airplane drifts after leveling off, nose just a little downwind. This will eliminate some of the sideload that may be placed on the wheels. However, the necessity for nosing downwind can be eliminated by gliding in with slightly less speed.

Make a 3-point landing, gliding at 120 mph with full flaps, or at 125 mph with ½ flaps.

Watch Brakes

On all landings, take particular care to avoid holding brakes while using rudder on the approach. Landing with brakes, or applying brakes before the full weight of the airplane settles, will cause blown tires and possible damage to the landing gear without the pilot ever knowing what is happening.

Caution

NEVER LAND WITH BRAKES. NEVER APPLY BRAKES BEFORE FULL WEIGHT OF PLANE SETTLES.

GO-AROUND

If the airplane is not on the ground within the first ⅓ of the runway, **go around again and make another approach and landing.**

1. Walk up throttles slowly. Copilot will check rpm before power is increased.

2. Retract flaps immediately after applying power. While they are being retracted raise nose slightly to overcome the loss of lift which occurs as flaps come fully up.

3. Copilot will call airspeed while flaps are retracting. Upon attaining an airspeed of 140 mph, reduce turbo regulators and throttles to desired setting. Signal copilot to reduce rpm. Copilot will then make an even adjustment on turbo regulators and synchronize the propellers.

Never reduce rpm before reducing manifold pressure. Remember: (1) for **reducing power,** reduce manifold pressure first, then reduce rpm; (2) for **increasing power,** increase rpm first, then increase manifold pressure.

If landing gear has been retracted, make a visual check before placing the landing gear switch in neutral. (Pilot will check and call aloud: "Up left"; and copilot: "Up right." Engineer will check tailwheel and report.)

Turn booster pumps "OFF" when leaving traffic above 1000 feet.

LANDING ROLL

After landing, use the entire runway for the landing roll, unless some emergency necessitates turning off at an intersection. Judgment of speed, and feel of the brakes, will tell you when and how to use them. After rolling half the runway, feel out the brakes by applying light pressure. If the braking effect is negligible, it means you will have to apply brakes sooner than normally in order to stop at a desired spot. If the brakes produce the desired slowing effect, you can leave them off until they are needed.

Finally, when brakes are used to slow down the airplane in order to turn off onto the taxi strip, apply them slowly and steadily, until you have attained a slow taxiing speed. Then turn off the runway. Use judgment in your application of brakes so that you will not have to apply additional power in order to turn onto the taxi strip.

AFTER-LANDING CHECK

While rolling down the runway, complete the after-landing section of the checklist from memory.

1. Copilot checks gage for proper hydraulic pressure (600 to 800 lb.).

2. Copilot opens and locks cowl flaps to cool engine and help slow down airplane.

3. Copilot turns turbos "OFF."

4. If no further takeoff is to be made, copilot turns booster pumps "OFF."

5. Copilot raises wing flaps upon signal from the pilot. Wing flaps are an aid in decreasing speed in the landing roll, and normally will be raised when speed is down to approximately 30 mph. When there is any possibility that flaps may be damaged by mud or slush, retract them immediately after contact with the ground.

6. Do not unlock the tailwheel before the end of the landing roll, except in emergency. (Tailwheel lock is operated by the copilot upon command from the pilot.)

7. Turn all generator switches to the "OFF" position.

End of Mission

After 30 seconds operation at 1200 rpm, signal the copilot to cut the inboard engines. Engines should not fire after mixture controls are in the "OFF" position. Advance throttles slowly so that the accelerating pump on the carburetor will not throw an extra charge into the cylinders and cause them to fire.

Turn off the runway and taxi toward the parking area. Be sure a crew man is out in front to guide you into the parking space.

Park the airplane with the tailwheel locked.

Be sure that chocks have been placed under the wheels before releasing the foot brakes.

Never set parking brakes upon return to the line. The hot brakes may cause the expander tubes to burst.

When the airplane is on the ramp, cut the outboard engines, after making sure that they are not fouled.

Contact tower by radio and report the airplane on the ramp.

Turn all electrical switches "OFF" before turning off master switch and battery switches. AC power switches must not be turned off until the engines have stopped and engine instruments have settled in neutral position. **Turn off main line and battery switches last.** This procedure will prevent arc-ing of relays, and eliminate heavy load on batteries when switches are turned on again.

Move the control column full forward, place rudder pedals in neutral, and raise the lock (in the floor to the right of the pilot's seat) to the "UP" position. Place the aileron lock in the control wheel.

Complete Form 1. Time ends when the airplane is in position on the ramp. Compute pilot time carefully. Make notations on Form 1A of things found wrong with the airplane; discuss the more serious items with the crew chief.

NIGHT FLYING

Don't fall for the belief, common among less experienced flyers, that "night flying is no different from day flying." Night flying is different from day flying. Your vision at night is different because you are using a different part of your eye (see **Physiology in Flight** and AAF Memorandum 25-5). Unless lights are properly grouped (as on runways) or easily identifiable (horizons, large cities, towns, etc.), your visual references are diminished considerably. Finally, when visibility is reduced and you have no clearly defined horizon, **night flying is instrument flying**.

Illusions in Night Flying

Night flying can be much more confusing than simple instrument flight through clouds. Probably many of the accidents and fatalities that occur in night flying result from the fact that pilots rely too much on their vision and other senses rather than on instruments. (See T.O. No. 30-100A-1.)

The inexperienced pilot is continually looking for some light on the ground by which he can orient himself. Unless he is flying near a large city where there are enough lights to make a good pattern, this practice of trying to orient himself in relation to the terrain is extremely hazardous. Many experienced pilots can tell how they have mistaken a star for a light beneath them, or how they thought lights were moving past, when actually their plane was turning about the lights.

The reason for the particular confusion in night flying is that a pilot's eyes may deceive him. He does not have any definite horizon to use as a plane of orientation; he has only isolated points of light. His sensation may tell him that these light-points are in a completely different relationship. As a result, when the airplane does not react as he expects it to, he becomes completely confused. In addition, the inexperienced pilot usually forgets his instruments and is so busy looking around that he glances at the instrument panel only after he has become confused and is already in a bad situation.

The only solution for this is to watch the **instrument panel** with only occasional glances out at the visual reference points. In night fly-

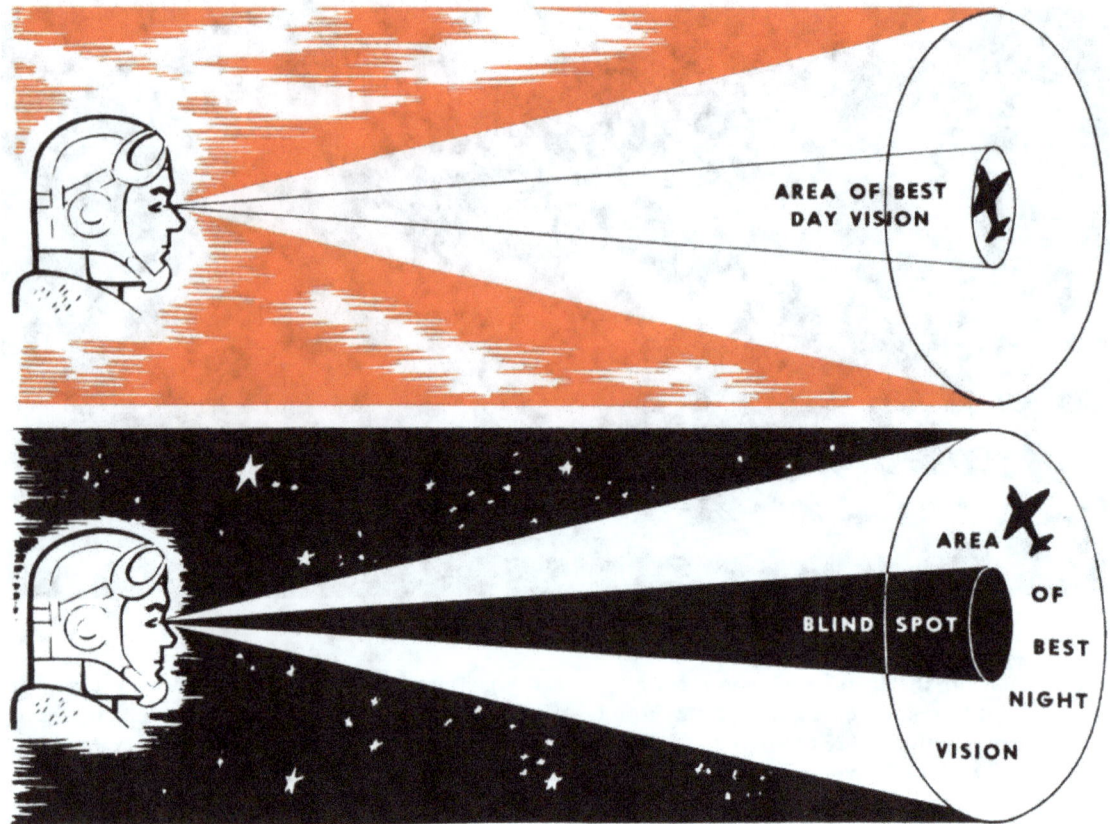

ing, use instruments as your major reference, and scattered lights only as a secondary reference.

Tips on Night Vision

Before flight don't subject your eyes to any bright lights: brightly lighted rooms, wing light beams, bright cockpit lights, etc.

Turn out all unnecessary cockpit lights; dim instrument panel lights.

Read instruments, maps, and charts rapidly; then look away. Use red light within the airplane whenever possible.

Lack of oxygen seriously impairs vision. At 12,000 feet without oxygen, for instance, night vision is only 50% efficient. Use oxygen from the ground up on all night flights to altitude.

Night Vision Precautions

Be sure that goggles, side windows, and wind shields are kept scrupulously clean. Scattered light on unclean surfaces reduces the contrast between faint lights and their background.

Be sure that all fluorescent lights, winglights, navigation lights, passing light, cockpit light, and individual instrument lights are in operating order.

Be sure that pilot, copilot, and engineer have individual flashlights.

Check radio operation and set proper frequencies. Your radio is especially important at night.

Know your field layout, the proper relationship of taxi strips to runways, etc. It's easy to become confused at night.

Takeoff

Obtain clearance from the tower before taxiing to the runway. Line up in the center of the runway and use runway lights for reference.

If visibility is poor and no horizon is visible, prepare to take off on instruments.

Maintain proper airspeed, but be sure you're climbing. It is imperative to hold a constant heading until you reach sufficient altitude for the turn.

Post observers at the side windows and top turret to give warning if you are turning into the path of other aircraft.

Remember that, for safety, 145 mph is the recommended climbing speed at night.

Don't start turns until you are at least 400 feet above the terrain. Don't reduce power until 200 feet altitude has been reached.

Night Landings

1. Fly compass headings on the various legs of the traffic pattern.

2. To line up properly with the runway and avoid overshooting or undershooting, begin a medium turn on the final approach **when the runway lights seem to separate.** On the downwind and base legs, the runway lights seem to be in a single row. As the airplane comes nearer to the runway on the base leg, the lights begin to separate into 2 rows. **This is the time to start the turn onto the approach.**

3. Avoid a low approach at night. Maintain constant glide, constant airspeed, and constant rate of descent by making slight changes in power and attitude.

4. Don't turn on wing lights while too high. They will become effective at 500 feet.

5. Don't try to sight down the wing light beam. Use the whole lighted area ahead and below for reference. Don't rely on winglights alone; use runway lights as a secondary reference. Winglights alone may induce you to level off for landing too late. Runway lights alone may cause you to level off too high, especially if there is haze or dust over the field.

6. If you are uncertain of your final approach, carry a little more power; this will prevent stalling out high. Carry power until you are sure of making contact with the ground. Avoid cutting power too high or too soon.

7. Check generators and batteries for proper operation. They carry a heavier load at night.

8. Check auxiliary power unit for operation in possible emergency. It should be on for all takeoffs and landing.

Taxiing Precautions

1. Keep use of landing lights while taxiing to a minimum; they burn out quickly. When taxiing use the winglights alternately as needed. This reduces the load that would be imposed on the electrical system by both lights. However, don't hesitate to use both lights if you really need them.

2. Make frequent checks of wheels and tires, using flashlights if landing gear inspection lights are not installed.

3. Using your flashlight, check cowling for signs of engine roughness.

4. When taxiing close to obstructions or parked aircraft, see that members of the ground crew walk ahead of each wing and direct taxiing by means of light signals.

5. Be particularly careful in judging distance from other taxiing aircraft. Sudden closure of distance is difficult to notice at night.

6. In case of failure or weakening of brakes, stop immediately and have the airplane towed in to the line. Faulty brakes are always hazardous. They are certain to cause accidents when taxiing at night.

COLD WEATHER OPERATION

Winter operation presents seasonal headaches in the operation and maintenance of the B-17 as it does in any other airplane.

Unless ground temperatures are below —23°C, no special procedures other than oil dilution are necessary so far as the pilot is concerned. Below this temperature, ground heaters must be used to preheat the engines and instruments, or warm hangar space must be available. (For detailed Winterization Instructions and Operation, see Technical Orders No. 01-20EF-51 and No. 00-60-3 and the PIF.)

Preheating.—Preheating is necessary under extreme cold weather conditions. Ground heaters are available for this purpose.

1. Cabin Compartment—When temperature is below —29°C it is necessary to heat the instrument panel to insure proper functioning of the instruments. Follow the prescribed procedure, utilizing openings in the nose and the bottom exit door.

2. Preheat propeller domes and engine front housing.

3. Preheat engine accessory section.

4. Thaw tailwheel assembly, if necessary.

5. Thaw control surface hinges.

Oxygen Equipment.—Operate all oxygen valves carefully in cold weather, opening and closing them slowly. (Rapid opening may cause sudden pressure and an explosion.)

Frost Prevention on Windows.—Leave opening in cockpit to permit air circulation, thus preventing frosting of windows.

Brakes.—When parking brakes have been in the "OFF" position for any length of time, the expander tubes stiffen in the contracted position. If pressure is applied suddenly, the expander tubes can be ruptured easily.

1. Have heat applied to brake drums if you think that the brakes are frozen.

2. Upon the first use of brakes, apply pressure gently. Do not apply full pressure until a number of light applications have been made.

3. While taxiing, if you suspect that moisture or water is present in the brakes, exercise the brakes more than usual. The extra heat thus generated will not only melt the snow or ice particles but will cause moisture to evaporate, leaving brakes practically dry. Be careful not to **overheat** brakes.

To Start Engines

1. Have the propellers pulled through at least 3 complete revolutions (9 blades). If difficulty is encountered in this operation, remove the lower spark plugs and clear the cylinders. Never back up the engine.

2. The proper use of engine primer will do much to facilitate engine starting in cold weather. Avoid overpriming, especially if first start is unsuccessful.

3. Leave mixture controls in "IDLE CUT-OFF" position.

4. Don't start on batteries if **auxiliary power** is available.

5. Start fuel booster pump.

6. Crack throttle to between 800 and 1000 rpm.

7. Set primer to engine being started and operate to expel air from lines.

8. As starter is meshed, operate primer with rapid full strokes to atomize the fuel. Limit hand priming to amount necessary to keep engine running.

9. When engine fires, place mixture control in "AUTO-RICH" position. In extreme cold weather, it may be necessary to stand by with hand primer for a short time to keep engine running.

10. Return primer to "OFF" (down) position after all engines are started. Never leave primer plunger in the up position when not priming engine. This allows fuel to pass directly to engine selected.

11. Don't use the starter repeatedly (4 or 5 times) without allowing it to cool off.

Over-priming

During a difficult starting, if the mixture controls have been moved from the "IDLE CUT-OFF" position, the engines will be over-primed. Fuel will flow from the blower-drain section of the engine.

1. Shut off the ignition switch. Place the throttle in the full open position. Have the propeller pulled through by hand to clear the engine of fuel.

2. Repeat starting procedure.

3. If several attempts at starting prove unsuccessful, locate cause of the trouble by consulting the **Handbook of Service Instructions.**

Oil Dilution After Starting

Follow the normal engine-starting procedure without regard for the oil dilution system. After starting, if oil pressure (a) is too high, (b) is fluctuating, or (c) drops as rpm is increased, this indicates that heavy viscous oil is in the system. **The condition can be corrected by pushing the oil dilution switches several times.**

Use this method with caution, and only when extreme weather conditions and lack of time do not permit normal engine warm-up. Remember that it is possible to cause engine failure by supplying the engine pump with pure gasoline.

If dilution is used during the warm-up, keep a close check on oil pressure throughout the warm-up and takeoff to guard against possible overdilution. (See p. 104.)

Warm-up

1. Keep flaps open for ground operation.

2. Check oil pressure. If there is no oil pressure indication within 30 seconds, shut down engine and investigate.

3. Idle engines at 900 to 1000 rpm until oil temperature rises to 40°C.

4. Operate turbo controls slowly through their entire range several times after engines have warmed up. Proper functioning of turbo regulator depends on flow of engine oil through the regulator. It is imperative that cold oil in the regulator be replaced by warm engine oil. Otherwise, closed waste gate and runway turbo may result on takeoff.

5. Operate propeller controls slowly through their entire range several times after engine oil reaches desired temperature. Proper propeller functioning depends on flow of engine oil through governor and propeller dome.

OIL DILUTION

Oil dilution is simply the introduction of gasoline into the engine oil supply to thin the oil, making starting of the engine easier in cold weather. Engine oil should be diluted before stopping the engines when it is suspected that the outside air temperature will drop below 5°C during the period the engine is to be stopped.

System—The system consists of 4 electric solenoid-operated oil dilution valves, each located on the front of its respective engine firewall. Four toggle switches are on the copilot's control panel. A fuel line from the carburetor connects to the Y oil drain valve, also a line from the dilution valve to the fuel pressure gage. Turning the oil dilution switch "ON" permits gasoline to flow through the valve into the oil line to circulate through the entire engine oil system. Engines must be running before dilution can be accomplished.

Procedure

1. Idle engines until oil temperature drops to approximately 40°C.

2. Run engines at 1000 to 1200 rpm for dilution.

3. Maintain oil temperatures at less than 50°C and an oil pressure above 15 lb. sq. in. If oil temperature rises above, or oil pressure falls below these limits, stop engines, allow to cool, and continue dilution.

4. If the airplane has an automatic dilution switch installed, simply dilute as instructed.

Anticipated Lowest Outside Temperature	Dilution Time in Minutes—One Period
4° to —12° C	2 minutes
—12° to —29° C	5 minutes
—29° to —46° C	7 minutes

5. If the plane has the manual dilution switch, hold the switch "ON" for the period shown in the table; release the switch only after engine has been stopped.

For each 5°C below —46°C, add one minute to the time given.

6. Proper operation of the dilution system is indicated by considerable drop in fuel pressure.

7. If it is necessary to service oil tanks, split the dilution period in half and service between the 2 periods.

8. Near the completion of the final dilution period, depress the propeller feathering button for 2 to 4 seconds, or a maximum change of 400 rpm, and pull out. Repeat several times.

9. Toward the end of final dilution period, operate supercharger regulator controls several times from low to high boost positions, taking 8 to 10 seconds for this procedure. This insures dilution of oil in turbo regulator. (The procedure is unnecessary on models equipped with electronic regulator.)

Whenever engines have been previously diluted and the airplane has not been flown, operate the engines for at least 30 minutes with oil temperature above 50°C before attempting **full dilution**. If a short ground run-up is made, the engines should be rediluted by reducing the dilution period accordingly. For instance, assume that dilution time is 5 minutes for full dilution. If engine run-up time is 15 minutes, the dilution time will be 2½ minutes.

After several days layover, during which time the engines have been started and diluted several times, it is advisable to ground-run the engines for 30 minutes at normal temperatures before takeoff to evaporate the gasoline in the oil.

Avoid **overdilution**. It causes the engine scavenging system to break down, resulting in a complete loss of all engine oil in a short time. Overdilution may be evidenced by a spewing of oil out of the engine breather, and a considerable drop in engine oil pressure.

CARBURETOR ICE PREVENTION IN FLIGHT

An understanding of the theory of carburetor ice formation and elimination is an important item in every pilot's technical knowledge. Tests have shown that under the most favorable conditions an engine will stop completely within 3 minutes after the first indications of icing have appeared. All B-17G and most B-17F airplanes are equipped with carburetor air temperature thermometers and gages to assist the pilot in detecting icing. The thermometers are so located that they measure inlet air temperature before fuel vaporization takes place.

Causes of Icing

The formation of carburetor ice is directly dependent upon the temperature and relative humidity of the carburetor inlet air. For ice to form, it is necessary that the carburetor inlet air temperature be less than 15°C and that the relative humidity be 50% or more. The ice forms in the adapter section, and sometimes at the fuel nozzle, as a result of the temperature drop (sometimes as much as 18°C) induced by fuel evaporation.

Icing caused by fuel evaporation does not affect the throttle valve of the Bendix Stromberg carburetor used on the B-17 airplane, as fuel is injected between the valve and the engine. Throttle icing can occur, however, under the following conditions:

1. Carburetor air inlet temperatures between 7°C and 10°C.
2. A relative humidity greater than 100%. (Rain drops or ice particles entering the air inlet.)
3. A throttle opening of less than 45°.

Icing is evidenced by rough engine operation, loss in manifold pressure, or abnormally high settings required of the throttle or turbo control levers to produce the desired manifold pressure.

Procedure for Ice Prevention

When operating under suspected icing conditions the following procedure is recommended:

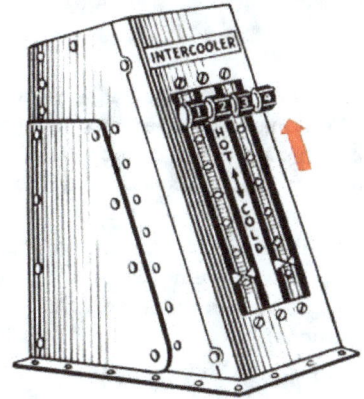

1. Carburetor air filters on and intercooler shutters full closed ("HOT") for taxiing, flight under 10,000 feet, and landing.

2. Above 10,000 feet, it may be necessary to open the shutters in proportion to the altitude to avoid exceeding 38°C carburetor air temperature. Filters may be left on up to 15,000 feet with full-throttle engine operation, but not higher because serious overspeeding of the turbos will result if they do not have overspeed control. (Under icing temperatures it is safe to keep filter on up to 15,000 feet.)

Emergency Ice Removal

To remove ice in an emergency use the following procedure:

1. Turn filter on if below 15,000 feet.

2. Close intercooler shutters, but do not exceed 38°C carburetor inlet air except momentarily.

3. Add up to 3″ of boost below 25,000 feet by reducing the throttle and increasing the turbo boost. Below 15,000 feet, more heat can be obtained by using full turbo and part throttle, not exceeding 34″ manifold pressure at 2200 rpm. Use this only for a short time, since excessive carburetor leanness results from the high carburetor deck pressures.

4. At or above 25,000 feet, close intercooler shutters only. Do not use filter or extra boost, because excessive turbo rpm will result unless turbos have overspeed control. Increase or decrease altitude to change the outside air temperature and reduce the amount of visible moisture such as fog, rain, snow, sleet, etc.

5. Generally, icing will not occur if the carburetor air inlet temperature is kept at 20°C or above.

If 38°C is not exceeded there will be no danger of detonation. It is more desirable to **prevent** formation of ice than to have to remove it in an emergency.

Experience proves that you can fly through severe icing conditions with normal precautions. However, a mild icing condition may cause the loss of an engine if you allow icing to progress to the point where corrective measures are ineffective.

ICING ON AIRCRAFT

Icing on aircraft in flight is a serious hazard. Ice accretion may occur at any temperature from near freezing down to more than —20°C when there is visible moisture in the atmosphere.

Avoid flying through icing zones when possible. Know how to remove ice when you do encounter it. Know your plane's limitations in icing conditions.

Ice on the Airplane

1. Reduces the efficiency of the airfoil, adds drag, and increases the stalling speed.

2. Makes your airplane difficult to control and maneuver.

3. Increases the drag of struts, fuselage, radio masts, landing gear, etc.

4. Increases the load.

5. Causes failure of or error in certain flight instruments.

Prepare for icing wherever there is visible moisture in the air at temperatures approaching or below the freezing level:

1. In freezing rains, in all frontal zones.

2. If there is sleet on the ground, somewhere aloft there is a layer of freezing rain, and above that a layer of air with temperature out of freezing range. Sleet itself is not considered too hazardous, although hail will cause immediate damage to wing and empennage leading edges, windshield and nose.

3. In cumulus clouds and others with vertical development, whenever they occur.

4. In orographic clouds, formed when moisture-laden air is forced upward over hills and mountain ranges.

5. Along fronts, in stratus and stratocumulus cloud formations.

Temperatures

Look for most severe icing when the temperature is between 0°C and —5°C. Icing may occur down to —20°C or colder. Low pressure areas on the airfoil may cause mild icing at temperatures a few degrees above freezing when other conditions are favorable to icing.

Propeller Anti-icer System

The propeller anti-icer system is designed to **prevent** the formation of ice on the propeller blades, **not to remove it.** Therefore, as a prerequisite to the satisfactory operation of the system, it is necessary to turn the propeller anti-icers "ON" upon encountering icing conditions and not after ice has formed.

Windshield Anti-icer System

The pilot's windshields are kept free from ice by the use of windshield wipers in conjunction with an alcohol spray. The controls for the system are on the sidewall above the pilot's control panel. Don't use the wipers on dry glass. This system also must be in operation **before** any ice has formed. It is designed to **prevent** ice, not to **remove** it.

De-icer System

The vacuum system of the B-17F operates both the de-icer boots and the flight instruments. Operation is obtained by means of 2 vacuum pumps mounted on the accessory case of the No. 2 and No. 3 engines. The system is so arranged that the pressure side of both pumps will inflate the de-icer boots while the vacuum side of one pump is operating to deflate them, and the vacuum side of the other pump is operating the flight instruments. In event of failure of either pump, use the remaining one to operate the instruments. That pump will also maintain the inflation of the boots. Their efficiency will be greatly reduced, however, because the boots have to depend upon their own elasticity for deflation.

The vacuum selector valve (control handle

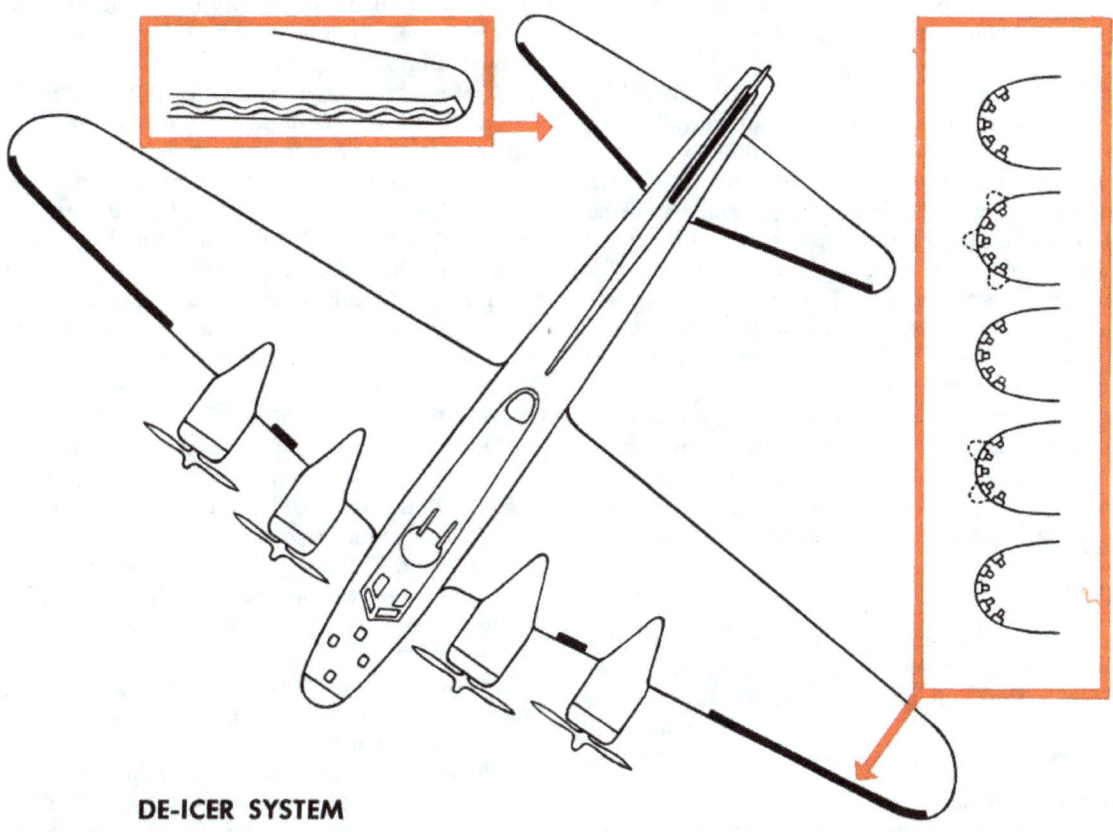

DE-ICER SYSTEM

at the pilot's left) directs the suction flow from the instruments to either the No. 2 or No. 3 vacuum pump. The de-icer control valve (control at left of pilot's seat) operates the de-icer distributor valve and also connects the pressure from both vacuum pumps and suction from one pump to the distributor valve. The de-icer distributor valve controls the alternate distribution of vacuum and pressure to the various de-icer boots.

Efficient performance of the de-icing system depends upon correct usage. It is generally considered good practice to permit the deposit of 1/8-inch ice on the boots before starting inflation. Then operate the boots intermittently as new ice is formed. The cycle of operation depends upon the severity of the icing conditions. Sometimes ice forms so rapidly that continuous operation of the boots is necessary. Watch continuous operation closely, as new ice may form over the cracked ice on the boots. The tubes then will pulsate ineffectively under a layer of ice.

Under conditions conducive to formation of smooth ice, it may be undesirable to use the de-icers. Glaze usually forms smooth layers of ice around the leading edges and conforms to their contour. Unless the ice becomes rough, the aerodynamic efficiency of the airfoil may not be greatly impaired. A ridge of ice left along the aft edges of the boots can have a more detrimental effect than the ice covering the entire leading edge.

Although the stalling speed changes only slightly, landing with de-icer boots operating is a poor policy. When a stall does occur with boots operating, it is more violent, and recovery requires a considerable increase in airspeed.

Pitot Heater

Before entering an icing condition turn the pitot heater switch on the pilot's control panel "ON." This will prevent the formation of ice in the pitot tube which would render the entire pitot system useless.

During Takeoff

Never take off with snow, ice or frost on the wings. Even loose snow cannot be depended upon to blow off, and a thin layer is sufficient to cause loss of lift and abnormal flying characteristics.

1. Where landing or taking off on a narrow strip of clear ice, crosswinds are particularly dangerous. Lack of traction causes loss of maneuverability. If the wind is gusty, the airplane may be blown completely off the icy runway before you can regain control.

2. If deep, heavy snow interferes with the takeoff but permits the airplane to be taxied, move slowly up and down the takeoff course several times to pack down a runway before attempting the actual takeoff. The depth and hardness of the snow determine whether takeoff or landing is practicable.

3. Regardless of outside temperature, always take off with cowl flaps open. The hazard of taking off with partially closed cowl flaps is too great, and the possibility of an engine cooling off excessively during the takeoff and rated power climb is negligible.

4. If necessary, you can take off immediately after oil dilution without the normal warm-up, provided that oil temperature is up, oil pressure is steady, and the engines are running smoothly. Cold oil properly diluted has the same viscosity as heated, undiluted oil, and therefore has the same ability to circulate and properly lubricate aircraft engines.

5. During takeoff the intercoolers may be turned on partially to prevent carburetor icing or to insure smooth engine operation.

During Flight and Landing

During flight, it is advisable to maintain cylinder-head temperatures not lower than 150°C. If the temperature drops below 125°C, rough operation may result.

During the glide, you should maintain the cylinder-head temperature at 125°C. Engine failure may result if the temperature drops below 100°C (212°F).

Make your approach and landing with the carburetor air filters on. This reduces the tendency toward carburetor icing.

During the approach for landing in cold weather, don't idle engines at low speed. They should be run up and checked frequently for ability to accelerate.

HOT WEATHER TIPS

Before Flight

1. Starting in hot weather requires less priming than starting under normal operating conditions.

2. Although outside air temperature is high, don't take off until oil temperature and oil pressure readings are normal.

3. Keep warm-up time and engine run-up time to a minimum. Run-up time for each engine should never exceed 30 seconds.

4. Always have cowl flaps open for all ground operation and for takeoff.

5. Remember, takeoff distances will be longer in hot weather.

6. While the airplane is on the ground, leave opening in fuselage for ventilation.

7. Use brakes as little as possible.

In Flight

1. Climb at not less than specified climbing speed (140 mph); lower climbing speeds will cause higher engine temperatures.

2. Low-altitude flying also will cause higher engine temperatures.

3. True stalling speed is greater in hot weather because the air is not as dense, but the indicated stalling speed is the same.

4. Don't expect to land in the same distance in hot weather as in cool weather. Indicated airspeed is the same, but groundspeed is increased because of the thinner air.

5. In hot weather, watch cylinder-head temperature closely and regulate it with cowl flaps.

OXYGEN

Your airplane was designed to operate just as well at high altitude as at low altitude.

Your body wasn't!

All organisms require oxygen to support life. At ground level you get plenty of oxygen from the surrounding air, which is packed down by the weight of the air above it.

As you go up there is less air above you. Therefore the air you breathe becomes thinner, your body is getting insufficient oxygen, and you begin to lose efficiency. At some altitude—varying with the individual—you'll become unconscious, and then, unless you get some extra oxygen quick . . . that's all, brother!

Remember, when the pressure of the air you're breathing is less than the normal atmospheric pressure of 10,000 feet, you need extra oxygen.

Therefore, your airplane has an oxygen system to meet the requirements of your body and allow you to function normally.

The equipment is excellent, simple to operate, and safe for flights up to 40,000 feet. **But it is not safe unless you understand it thoroughly and strictly observe the rules regarding its use.** You can't take short cuts with oxygen and live to tell about it!

The lack of oxygen, known as anoxia, gives no warning. If it hits you, you won't know it until your mates revive you from unconsciousness, if they can. Therefore, you must check the condition and operation of your equipment with extreme care, and continue to check it regularly as often as possible during flight.

Your oxygen mask is an item of personal issue. Take care of it. It's as important as your life.

Before you use the mask in flight, have it fitted carefully by your personal equipment officer, or his qualified assistants. They will see that you have the right size, that it fits perfectly, and that the studs to hold it are properly fixed to your helmet.

Bring it in for re-checking whenever necessary. The straps will stretch slightly after a period of use. It's a good idea to have the fit re-checked regularly whether you think it needs it or not.

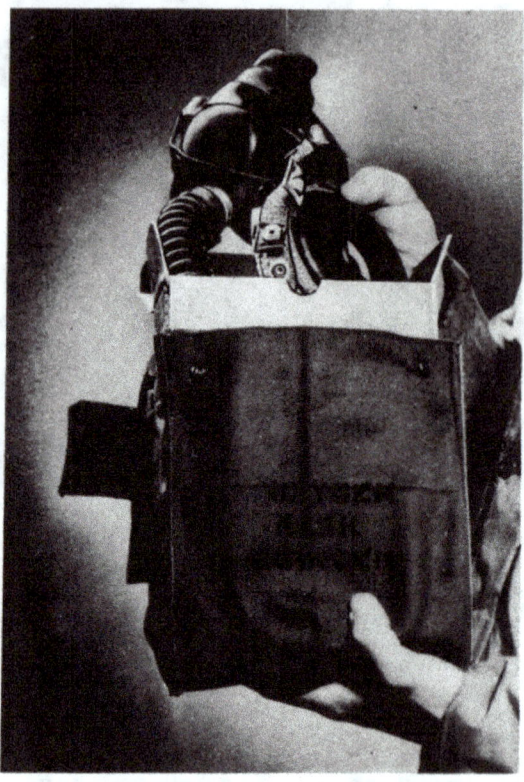

Keep your mask in kit when not in use.

Draw the mask before each mission. Return it to the supply room afterward. Equipment personnel will check it for repair and cleaning. But don't assume that this procedure relieves you of the responsibility of your own regular inspection and care of the mask.

Before each mission, make the following checks on your mask:

1. Look the mask and helmet over carefully for worn spots or worn straps, loose studs, or evidence of deterioration in facepiece and hose.

2. Put the mask on carefully. Slip the edges of the facepiece under the helmet. Adjust the straps, if necessary, to get a good fit.

3. Test for leak. Hold your thumb over the end of the hose and breathe in **gently**. The

mask should collapse on your face, with no air entering. Don't inhale strongly because the mask would seal anyway in that case, even with a leak.

Clip hose in position to allow full head movement

4. Clip the end of the regulator hose to your jacket in such a position that you can move your head around fully without twisting or kinking the mask hose or pulling on the mask hose or pulling on the quick-disconnect. Get the personal equipment section to sew a tab on your jacket at the proper spot.

5. See that the gasket is properly seated on the male end of the quick-disconnect fitting between mask and regulator hose. Plug in the fitting and **test the pull.** If it comes apart easily, spread the prongs with a knife blade.
Note: **This is only a temporary adjustment. As soon as possible report the difficulty to the equipment men and let them replace the fitting if necessary.**

General Tips: Vapor in your breath will freeze in the mask at extremely low temperatures. If you detect freezing, squeeze the mask to prevent ice particles from clogging the oxygen inlet.

Don't let anyone else wear your mask except in emergencies.

Keep it in the kit between flights, and keep it clean.

Report anything wrong with the functioning or condition of the mask when you turn it in after a flight.

Oxygen Regulator

A demand regulator is mounted at each station in the plane. There are 2 types of demand regulators, the Airco and the Pioneer. You may find either one on your plane. They look slightly different but the principle of operation is the same in both.

A demand regulator is one that furnishes oxygen on demand, or only when you inhale. It furnishes no oxygen when you exhale. Obviously, this is a more economical principle than a continuous flow of oxygen.

Pioneer Demand Regulator

Airco Demand Regulator. Auto-mix "ON"

Airco Demand Regulator. Auto-mix "OFF"

The regulator has an auto-mix mechanism controlled by a lever on the side of the cover. **The lever should be in the "ON" position at all times** when the system is in use (except in certain emergencies). When the lever is in the "ON" position, oxygen furnished below 30,000 feet is mixed with air. The mixture is controlled automatically by an aneroid to furnish the correct amount of oxygen which your body requires for a given altitude. Above approximately 30,000 feet the air inlet closes and you get 100% oxygen, although the lever in the regulator is still in the "ON" position.

With the lever in the "OFF" position, 100% oxygen is furnished. This operation wastes oxygen.

When breathing oxygen, **never turn the lever to "OFF" except in the following cases:**

1. To give 100% oxygen to a wounded man below 30,000 feet.

2. If poison gas is present in plane.

Operation of Emergency Valve

3. If the airplane commander prescribes breathing 100% oxygen all the way up to high altitude as a protection against the bends.

To operate the emergency valve, turn the red knob on the intake side of the regulator in the direction indicated on the regulator face. **Caution: Never pinch the mask hose or block the oxygen flow when the emergency valve is turned to "ON." This action breaks the regulator diaphragm.**

Turning emergency valve to "ON" causes the oxygen flow to bypass the demand mechanism and to flow continuously into the mask. It is extremely **wasteful of oxygen.** Leaving the valve "ON" bleeds the whole airplane oxygen system in a short time.

Never turn the emergency valve to "ON," except:

1. To revive a crew member.

2. In cases of excessive mask leakage.

3. When you have to take off your mask temporarily, for example, to blow your nose, vomit, or spit. In those cases unhook one side of the mask and hold it as close to your face as possible.

Make the following checks before each flight:

Check tightness of knurled collar

1. Check the tightness of the knurled collar. It should be so tight that the movement of the regulator hose will not turn the elbow.

2. Open the emergency valve slightly and see that there is a flow of oxygen. Be sure to close it tight again.

OXYGEN PANEL LOCATED AT EACH STATION

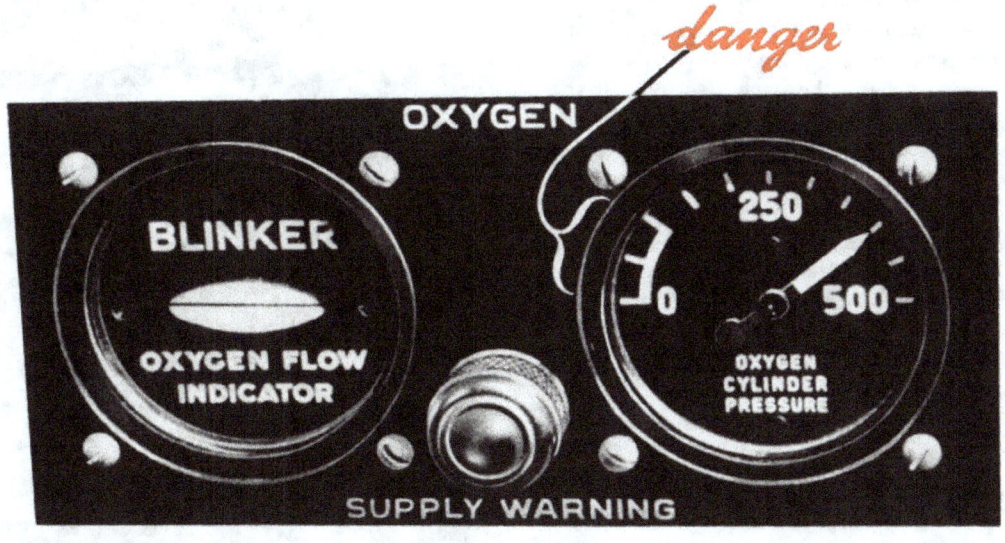

Flow indicator. Warning light. Pressure gage.

Flow Indicator

The flow indicator on the oxygen panel winks open and shut as the oxygen flows. The blinker may not operate normally at ground level with the auto-mix lever at "ON." Therefore, **before the flight plug in your mask, turn the auto-mix lever to "OFF" and see that the blinker works as you breathe.**

Be sure to move the lever back to "ON" before flight.

The blinker does not work when the emergency valve is "ON."

(Note: Some models have ball-in-cylinder flow indicators.)

Watch your flow indicator during flight. It is the only indication you have that the oxygen is flowing regularly. If it stops blinking (or if ball stops bouncing), use your portable equipment and plug in at another station if possible.

Blinker flow indicator, open

Blinker flow indicator, closed

Pressure Gage and Warning Light

Before flight, check the pressure gage on your panel. When the system on your plane is full the pressure should be between 400 and 425 lb. sq. in. Check the gage also against the gages at other stations. There may be some variation between stations because of different tolerances in the gages, but if yours is more than 50 lb. sq. in. off the others, investigate.

When the pressure gets down to between 95 and 105 lb. sq. in., the amber warning light in the center of the panel goes on. That means you haven't much of your oxygen supply left. When your light goes on emergency action is necessary.

The regulator does not work properly when the pressure gets below 50 lb. sq. in. If you can't get downstairs at that time, use your portable equipment until you can descend.

Walk-around Equipment

All stations have the small green type A-4 cylinder, equipped with gages and regulators. The regulators furnish 100% oxygen on demand.

Before each flight, check to see that your walk-around bottle is within easy reach. Look at the gage. If the pressure is more than 50 lb. sq. in. under the pressure of the airplane system, recharge the bottle.

There is a recharging hose at each station. Snap the hose fitting on the nipple of the regulator. Push it home until it clicks and locks. When the bottle has filled to the pressure of the plane system, turn the hose clamp clockwise and remove hose fitting. You can carry out this operation while your mask is plugged into the bottle you are filling.

Always use a walk-around bottle if you have to disconnect from the airplane system. Hold your breath while you are switching to the bottle. Clip the A-4 bottle to your jacket.

The duration of the walk-around oxygen supply is variable—usually only 5 minutes. Don't depend on it to last very long, regardless of what you have heard about the capacity.

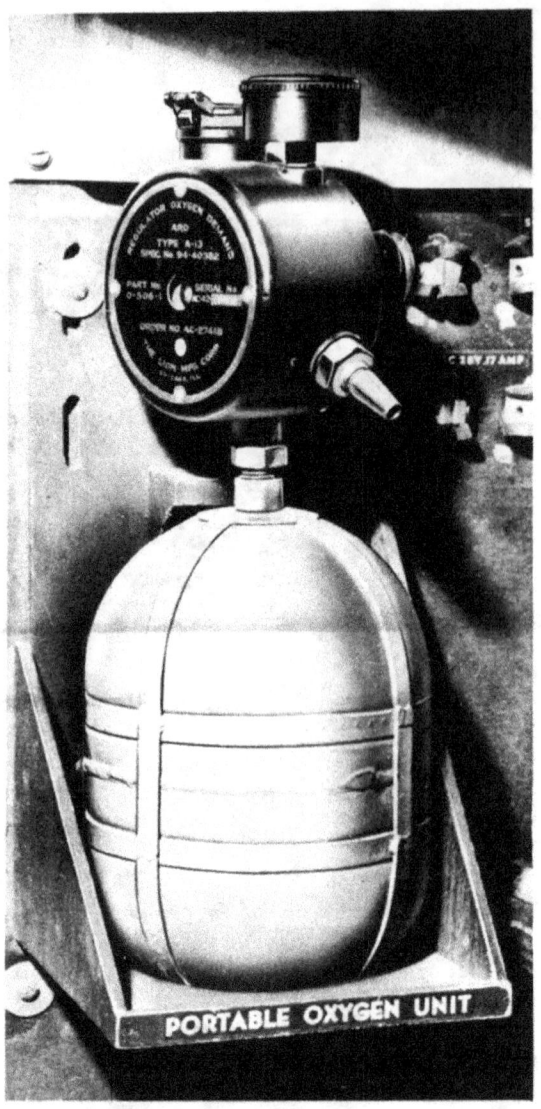

A-4 walk-around bottle on bracket at airplane station

Keep watching the gage, and recharge the bottle when it needs it.

Always recharge walk-around equipment after use.

Bailout Cylinders

The bailout cylinder is a small high-pressure oxygen cylinder, with a gage attached, which furnishes a continuous flow of oxygen.

The cylinder comes in a heavy canvas pocket provided with tying straps. Have this pocket sewed and tied securely to the harness of your parachute.

Before flight, check to see that the pressure of the cylinder is at 1800 lb. sq. in. If you have to bail out at a high altitude, securely plug the bayonet connection on the hose into the adapter on your mask, open the valve, and then disconnect your mask from the plane supply.

In using duration charts, figure in only the cylinders remaining intact, if any of them have been shot out.

Remember YOUR OXYGEN EQUIPMENT IS YOUR LIFE

Keep bailout cylinder hose plugged into mask

MAN HOURS OF AVAILABLE OXYGEN

BLACK FIGURES INDICATE AUTO-MIX "ON" **RED FIGURES INDICATE AUTO-MIX "OFF"**

CAUTION—The auto-mix in the "OFF" position rapidly diminishes the available oxygen supply. Do not use this position unless it is necessary to get *pure oxygen!*

AIRCO REGULATORS
TYPE A-12

GROUP I (5 G-1 Cylinders) — Pilot, Navigator and Top Turret Filler

Gage Pres. / Alt. Ft.	400	350	300	250	200	150	100	50	
40,000	41.5 / 41.5	35.6 / 35.6	29.4 / 29.4	23.6 / 23.6	17.8 / 17.8	12.0 / 12.0	5.8 / 5.8		E
35,000	29.5 / 29.5	25.3 / 25.3	20.9 / 20.9	16.8 / 16.8	12.6 / 12.6	8.5 / 8.5	4.0 / 4.0		M
30,000	21.5 / 22.0	18.5 / 18.9	15.2 / 15.6	12.2 / 12.5	9.2 / 10.4	6.0 / 6.2	3.0 / 3.0		E
25,000	16.5 / 21.0	14.1 / 18.0	11.5 / 14.9	9.0 / 11.9	7.0 / 9.0	4.7 / 6.0	2.0 / 2.9		R
20,000	13.0 / 23.5	11.1 / 20.2	9.2 / 16.6	7.4 / 13.3	5.5 / 10.1	3.7 / 6.8	1.5 / 3.2		G
15,000	10.0 / 28.5	8.6 / 24.5	7.0 / 20.2	5.7 / 16.2	4.0 / 12.2	3.9 / 8.2	1.4 / 3.9		E
10,000	8.0 / 48.5	6.8 / 41.7	5.6 / 34.4	4.5 / 27.6	3.4 / 20.8	2.3 / 14.0	1.1 / 6.7		N
5,000	6.5 / —	5.5 / —	4.6 / —	3.7 / —	2.8 / —	1.8 / —	1.0 / —		C
S. L.	5.5 / —	4.7 / —	3.9 / —	3.1 / —	2.3 / —	1.5 / —	0.7 / —		Y

GROUP II (4 G-1 Cylinders) — Copilot, Bombardier and Top Gunner

Gage Pres. / Alt. Ft.	400	350	300	250	200	150	100	50	
40,000	33.2 / 33.2	28.6 / 28.5	23.6 / 23.6	19.0 / 18.9	14.2 / 14.2	9.6 / 9.6	4.6 / 4.6		E
35,000	23.6 / 23.6	20.3 / 20.3	16.8 / 16.7	13.4 / 13.4	10.2 / 10.1	6.8 / 6.8	3.4 / 3.3		M
30,000	17.2 / 17.6	14.8 / 15.1	12.2 / 12.5	9.8 / 10.0	7.4 / 7.6	5.0 / 5.0	2.4 / 2.4		E
25,000	13.2 / 16.8	11.2 / 14.4	9.2 / 11.9	7.4 / 9.6	5.6 / 7.2	3.8 / 4.8	1.8 / 3.3		R
20,000	10.4 / 18.8	9.0 / 16.2	7.4 / 13.3	6.0 / 10.7	4.4 / 8.1	3.0 / 5.4	1.4 / 2.6		G
15,000	8.0 / 22.8	6.8 / 19.6	5.6 / 16.2	4.6 / 13.0	3.4 / 9.9	2.4 / 6.6	1.2 / 3.2		E
10,000	6.4 / 38.8	5.4 / 33.4	4.6 / 27.5	3.6 / 22.1	2.8 / 16.7	1.8 / 11.2	0.8 / 5.4		N
5,000	5.2 / —	4.4 / —	3.6 / —	3.0 / —	2.2 / —	1.4 / —	0.8 / —		C
S. L.	4.4 / —	3.8 / —	3.2 / —	2.4 / —	1.8 / —	1.2 / —	0.6 / —		Y

PIONEER REGULATORS
TYPE A-12

GROUP I (5 G-1 Cylinders)

Gage Pres. / Alt. Ft.	400	350	300	250	200	150	100	50	
40,000	41.5 / 41.5	35.6 / 35.6	29.4 / 29.4	23.6 / 23.6	17.8 / 17.8	12.0 / 12.0	5.8 / 5.8		E
35,000	29.5 / 30.0	25.3 / 25.8	20.9 / 21.3	16.8 / 17.1	12.6 / 12.9	8.5 / 8.7	4.0 / 4.2		M
30,000	21.5 / 22.5	18.5 / 19.3	15.2 / 15.9	12.2 / 12.8	9.2 / 9.6	6.0 / 6.5	3.0 / 3.1		E
25,000	16.5 / 22.0	14.1 / 18.4	11.5 / 15.6	9.0 / 12.5	7.0 / 9.4	4.7 / 6.3	2.0 / 3.0		R
20,000	13.0 / 39.0	11.1 / 33.5	9.2 / 26.6	7.4 / 22.2	5.5 / 16.7	3.7 / 11.3	1.5 / 5.4		G
15,000	10.0 / 38.0	8.6 / 32.6	7.0 / 26.9	5.7 / 21.6	4.0 / 16.3	3.9 / 11.0	1.4 / 5.3		E
10,000	8.0 / 37.5	6.8 / 32.2	5.6 / 26.6	4.5 / 21.3	3.4 / 16.1	2.3 / 10.8	1.1 / 5.2		N
5,000	6.5 / 28.5	5.5 / 24.5	4.6 / 20.2	3.7 / 16.1	2.8 / 12.2	1.8 / 8.2	1.0 / 3.9		C
S. L.	5.5 / 30.0	4.7 / 25.8	3.9 / 21.3	3.1 / 17.1	2.3 / 12.9	1.5 / 8.7	0.7 / 4.2		Y

GROUP II (4 G-1 Cylinders)

Gage Pres. / Alt. Ft.	400	350	300	250	200	150	100	50	
40,000	33.2 / 33.2	28.6 / 28.5	23.6 / 23.6	19.0 / 18.9	14.2 / 14.2	9.6 / 9.6	4.6 / 4.6		E
35,000	23.6 / 24.0	20.2 / 20.6	16.8 / 19.0	13.4 / 13.7	10.2 / 10.3	6.8 / 6.9	3.4 / 3.3		M
30,000	17.2 / 18.0	14.8 / 15.5	12.2 / 12.8	9.8 / 10.2	7.4 / 7.7	5.0 / 5.2	2.4 / 2.5		E
25,000	13.2 / 17.6	11.2 / 14.7	9.2 / 12.5	7.4 / 10.0	5.6 / 7.6	3.8 / 7.1	1.8 / 2.4		R
20,000	10.4 / 31.2	9.0 / 26.8	7.4 / 22.1	6.0 / 17.8	4.4 / 13.4	3.0 / 9.0	1.4 / 4.3		G
15,000	8.0 / 30.4	6.8 / 26.1	5.6 / 21.6	4.6 / 17.3	3.4 / 13.0	2.4 / 8.8	1.2 / 4.2		E
10,000	6.4 / 30.0	5.4 / 25.9	4.6 / 21.3	3.6 / 17.1	2.8 / 12.9	1.8 / 8.7	0.8 / 4.2		N
5,000	5.2 / 22.8	4.4 / 19.6	3.6 / 16.2	3.0 / 13.0	2.2 / 9.8	1.4 / 6.6	0.8 / 3.1		C
S. L.	4.4 / 24.0	3.8 / 20.6	3.2 / 17.0	2.4 / 13.7	1.8 / 10.3	1.2 / 7.0	0.6 / 3.3		Y

RESTRICTED

MAN HOURS OF AVAILABLE OXYGEN

BLACK FIGURES INDICATE AUTO-MIX "ON"   **RED FIGURES INDICATE AUTO-MIX "OFF"**

NOTE: Each turret cylinder, Type F-1, will supply one man for approximately 2 hours at 30,000 feet, 2½ hours at 25,000 feet, 3 hours at 20,000 feet

AIRCO REGULATORS — TYPE A-12

GROUP III (6 G-1 Cylinders) — Bomb Bay, Radio Operator, Side Gunner, Tail Gunner, and Ball Turret Filler

Gage Pres. Alt. Ft.	400	350	300	250	200	150	100	50
40,000	49.8 / 49.8	42.8 / 42.8	35.4 / 35.4	28.4 / 28.4	21.4 / 21.2	14.4 / 14.4	7.0 / 6.9	E
35,000	35.4 / 35.4	30.4 / 30.4	25.0 / 25.0	20.2 / 20.1	15.2 / 15.1	10.2 / 10.2	5.0 / 4.9	M
30,000	25.8 / 26.4	22.2 / 22.6	18.2 / 18.7	15.6 / 15.0	11.0 / 11.3	7.4 / 7.5	2.8 / 3.6	E
25,000	19.8 / 25.2	16.8 / 21.6	13.8 / 17.8	11.2 / 14.3	8.4 / 10.8	5.6 / 7.2	2.8 / 3.4	R
20,000	15.6 / 28.2	13.6 / 24.2	11.0 / 19.9	8.8 / 16.0	6.6 / 12.1	4.4 / 8.1	2.2 / 3.9	G
15,000	12.0 / 34.2	10.4 / 29.4	8.6 / 24.2	6.8 / 19.4	5.2 / 14.7	3.4 / 9.8	1.6 / 4.7	E
10,000	9.6 / 58.2	8.2 / 50.0	6.8 / 41.2	5.4 / 33.1	4.2 / 25.0	2.8 / 16.8	1.4 / 8.1	N
5,000	7.8 / —	6.6 / —	5.6 / —	4.2 / —	3.4 / —	2.2 / —	1.2 / —	C
S.L.	6.6 / —	5.6 / —	4.6 / —	3.8 / —	2.8 / —	1.8 / —	0.8 / —	Y

GROUP IV (3 G-1 Cylinders) — Radio Compartment (2 Outlets), Side Gunner and Tail Gunner

Gage Pres. Alt. Ft.	400	350	300	250	200	150	100	50
40,000	24.9 / 24.9	21.4 / 21.4	17.7 / 17.7	14.2 / 14.2	10.7 / 10.7	7.2 / 7.2	3.5 / 3.5	E
35,000	17.7 / 17.7	15.2 / 15.2	12.5 / 12.5	10.1 / 10.1	7.6 / 7.6	5.1 / 5.1	2.5 / 2.5	M
30,000	12.9 / 13.2	11.1 / 11.3	9.1 / 9.4	7.3 / 7.5	5.5 / 5.7	3.7 / 3.8	1.4 / 1.8	E
25,000	9.9 / 12.6	8.4 / 10.8	6.9 / 8.9	5.6 / 7.2	4.2 / 5.4	2.8 / 3.6	1.4 / 1.7	R
20,000	7.8 / 14.1	6.8 / 12.1	5.5 / 10.0	4.4 / 8.0	3.3 / 6.1	2.2 / 4.1	1.1 / 1.9	G
15,000	6.0 / 17.1	5.2 / 14.7	4.3 / 12.1	3.4 / 9.7	2.6 / 7.3	1.7 / 4.9	0.8 / 2.4	E
10,000	4.8 / 29.1	4.1 / 25.0	3.4 / 20.5	2.7 / 16.6	2.1 / 12.3	1.4 / 8.4	0.7 / 4.0	N
5,000	3.9 / —	3.3 / —	2.8 / —	2.1 / —	1.7 / —	1.1 / —	0.6 / —	C
S.L.	3.3 / —	2.8 / —	2.3 / —	1.9 / —	1.4 / —	0.9 / —	0.4 / —	Y

PIONEER REGULATORS — TYPE A-12

GROUP III (6 G-1 Cylinders)

Gage Pres. Alt. Ft.	400	350	300	250	200	150	100	50
40,000	49.8 / 49.8	42.8 / 42.8	35.4 / 35.4	28.4 / 28.4	21.4 / 21.3	14.4 / 14.4	7.0 / 6.9	E
35,000	35.4 / 36.0	30.4 / 30.9	25.0 / 25.5	20.2 / 20.5	15.2 / 15.4	10.2 / 10.4	5.0 / 5.0	M
30,000	25.8 / 27.0	22.2 / 23.2	18.2 / 19.1	15.6 / 15.3	11.0 / 11.5	7.4 / 7.8	2.8 / 3.7	E
25,000	19.8 / 26.4	16.8 / 22.0	13.8 / 18.7	11.2 / 15.0	8.4 / 11.3	5.6 / 7.6	2.8 / 3.8	R
20,000	15.6 / 46.8	13.6 / 40.2	11.0 / 33.1	8.8 / 26.6	6.6 / 20.1	4.4 / 13.5	2.2 / 6.5	G
15,000	12.0 / 45.6	10.4 / 39.1	8.6 / 31.7	6.8 / 25.9	5.2 / 19.5	3.4 / 13.2	1.6 / 6.3	E
10,000	9.6 / 45.0	8.2 / 38.7	6.8 / 31.9	5.4 / 25.6	4.2 / 19.3	2.8 / 13.0	1.4 / 6.3	N
5,000	7.8 / 32.2	6.6 / 29.4	5.6 / 24.2	4.2 / 19.4	3.4 / 14.7	2.2 / 9.9	1.2 / 4.5	C
S.L.	6.6 / 36.0	5.6 / 31.9	4.6 / 25.5	3.8 / 20.5	2.8 / 15.4	1.8 / 10.4	0.8 / 5.0	Y

GROUP IV (3 G-1 Cylinders)

Gage Pres. Alt. Ft.	400	350	300	250	200	150	100	50
40,000	24.9 / 24.9	21.4 / 21.4	17.7 / 17.7	14.2 / 14.2	10.7 / 10.7	7.2 / 7.2	3.5 / 3.5	E
35,000	17.7 / 18.0	15.2 / 15.5	12.5 / 12.8	10.1 / 10.3	7.6 / 7.7	5.1 / 5.2	2.5 / 2.5	M
30,000	12.9 / 13.5	11.1 / 11.6	9.1 / 9.6	7.3 / 7.7	5.5 / 5.8	3.7 / 3.9	1.8 / 1.9	E
25,000	9.9 / 13.2	8.4 / 11.0	6.9 / 9.4	5.6 / 7.5	4.2 / 5.7	2.8 / 3.8	1.4 / 1.8	R
20,000	7.8 / 23.4	6.8 / 20.1	5.4 / 16.6	4.4 / 13.3	3.3 / 10.0	2.2 / 6.8	1.1 / 3.3	G
15,000	6.0 / 22.8	5.2 / 19.6	4.3 / 16.2	3.4 / 13.0	2.6 / 9.8	1.7 / 6.6	0.8 / 3.2	E
10,000	4.8 / 22.5	4.1 / 19.3	3.4 / 16.0	2.7 / 12.8	2.1 / 9.7	1.4 / 6.5	0.7 / 3.1	N
5,000	3.9 / 16.1	3.3 / 14.7	2.8 / 12.1	2.1 / 9.7	1.7 / 7.3	1.1 / 4.9	0.6 / 2.3	C
S.L.	3.3 / 18.0	2.8 / 15.5	2.3 / 12.8	1.9 / 10.3	1.4 / 7.7	0.9 / 5.2	0.4 / 2.5	Y

OXYGEN SYSTEM

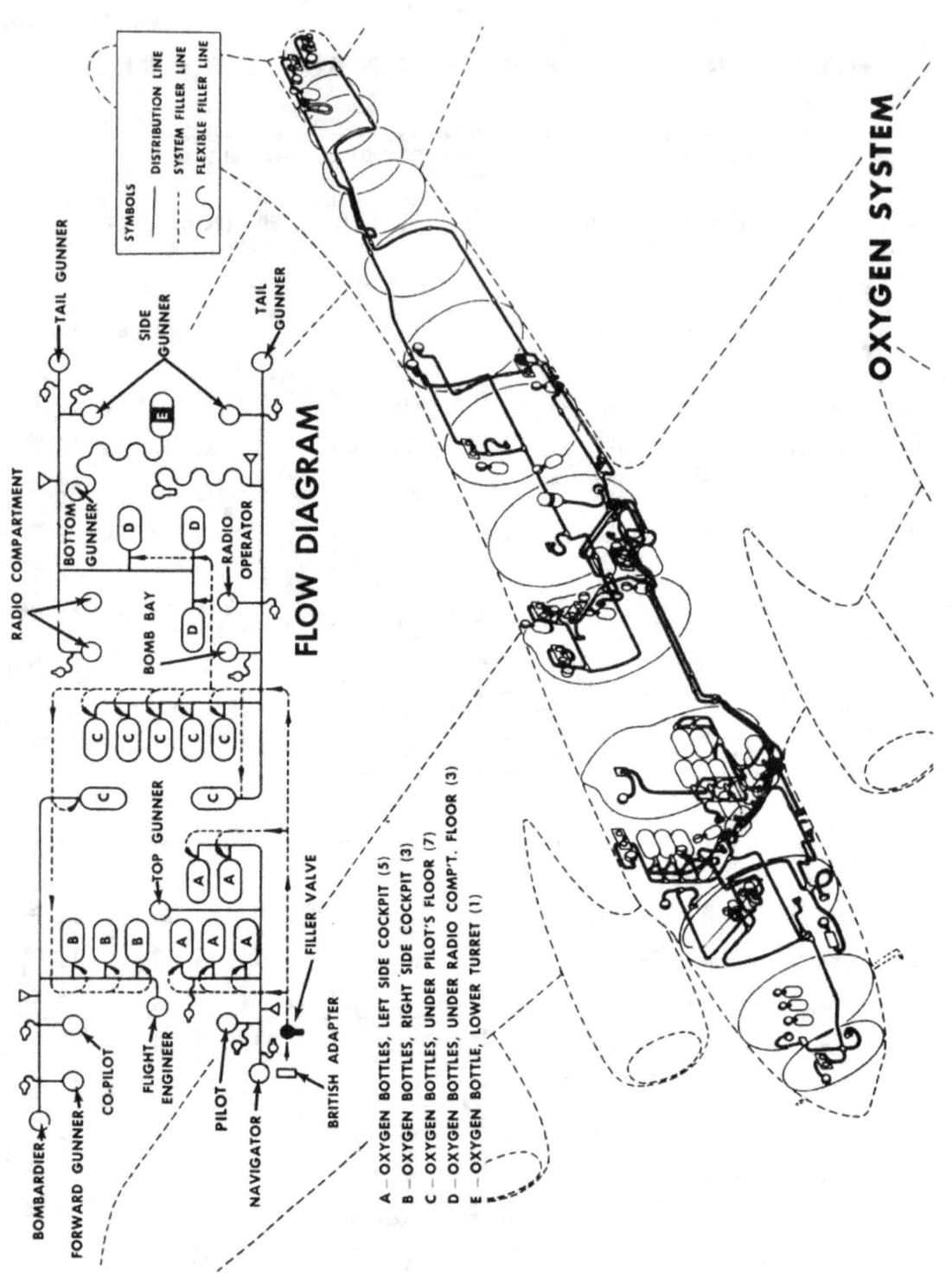

FLOW DIAGRAM

SYMBOLS
- ——— DISTRIBUTION LINE
- – – – SYSTEM FILLER LINE
- ∿∿∿ FLEXIBLE FILLER LINE

A — OXYGEN BOTTLES, LEFT SIDE COCKPIT (5)
B — OXYGEN BOTTLES, RIGHT SIDE COCKPIT (3)
C — OXYGEN BOTTLES, UNDER PILOT'S FLOOR (7)
D — OXYGEN BOTTLES, UNDER RADIO COMP'T. FLOOR (3)
E — OXYGEN BOTTLE, LOWER TURRET (1)

FORMATION

When you get into combat you will learn that your best assurance of becoming a veteran of World War II is the good, well-planned, and well-executed formation.

Formation flying is the first requisite of successful operation of the heavy bomber in combat. Groups that are noted for their proficiency in formation flying are usually the groups with the lowest casualty rates. Proper formation provides: controlled and concentrated firepower, maneuverability, cross-cover, precise bombing pattern, better fighter protection.

Heavy Bomber Formations

Formation flying in 4-engine airplanes presents greater problems than formation flying in smaller aircraft. The problems increase in almost direct proportion to the airplane's size and weight. In the B-17, relatively slower response to power and control changes require a much higher degree of **anticipation** on the part of the pilot. Therefore you must allow a greater factor of safety.

Violent maneuvers are unnecessary and seldom encountered. Close flying becomes an added hazard which accomplishes no purpose and is not even an indication of a good formation. Bear in mind that it is much more difficult to maintain position when flying with proper spacing between airplanes than with wings overlapping.

Safety first is a prerequisite of a good formation because a greater number of lives and a larger amount of equipment is in the hands of the responsible pilot in a large 4-engine airplane.

Clearance

In flying the Vee formation, aircraft will not be flown closer to one another than 50 feet from nose to tail and wingtip to wingtip. Maintain this horizontal clearance whenever vertical clearance is less than 50 feet, thus providing a minimum of 50 feet clearance between wingtips as well as the line of nose and tail under all formation flying conditions.

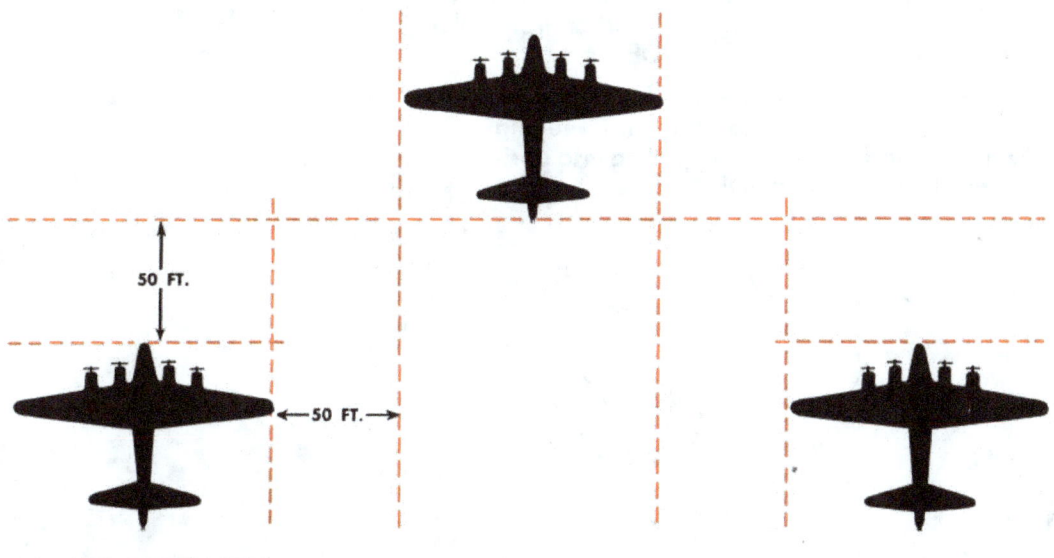

THE VEE FORMATION

RESTRICTED
FORMATION TAKEOFFS

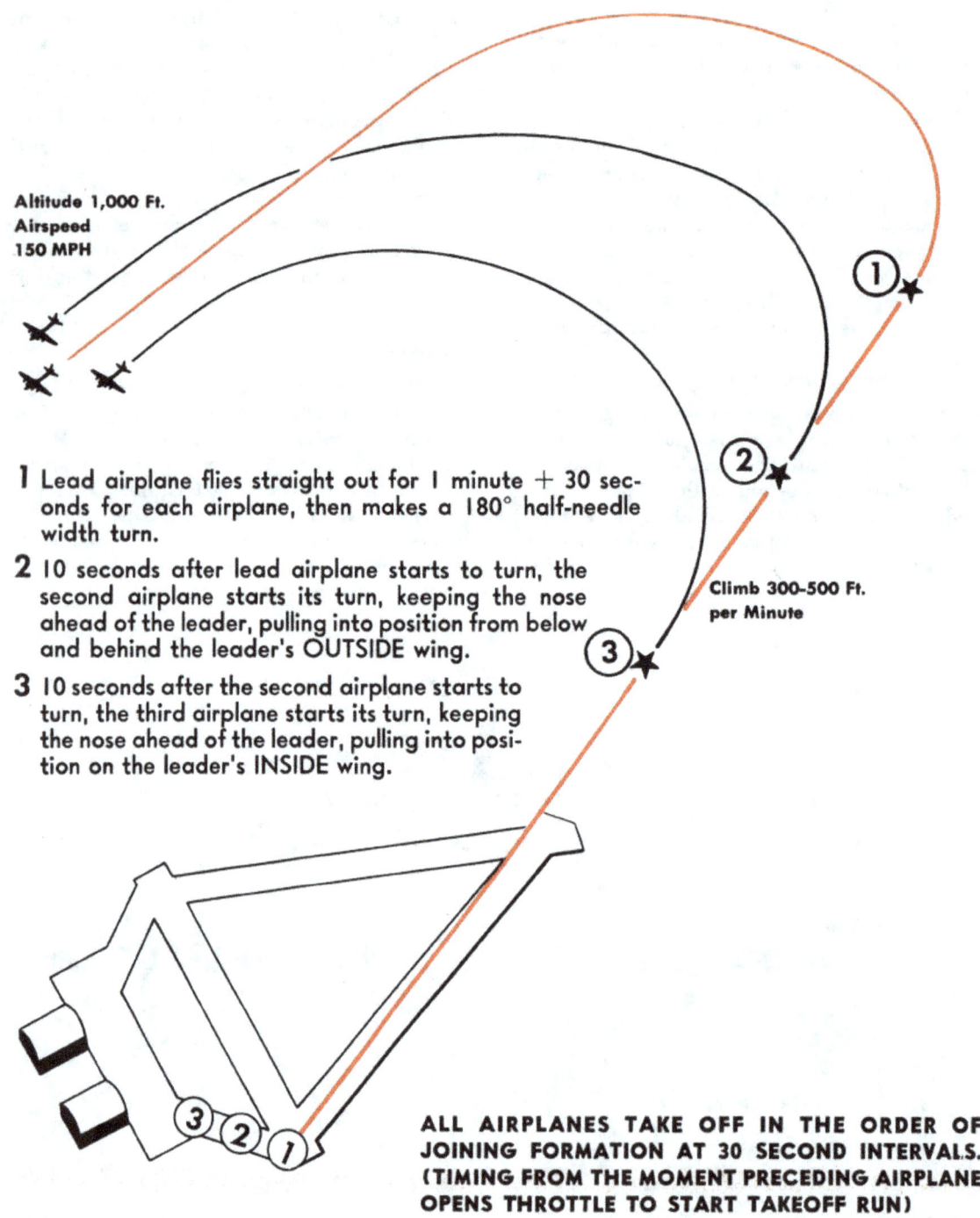

Altitude 1,000 Ft.
Airspeed 150 MPH

Climb 300-500 Ft. per Minute

1. Lead airplane flies straight out for 1 minute + 30 seconds for each airplane, then makes a 180° half-needle width turn.
2. 10 seconds after lead airplane starts to turn, the second airplane starts its turn, keeping the nose ahead of the leader, pulling into position from below and behind the leader's OUTSIDE wing.
3. 10 seconds after the second airplane starts to turn, the third airplane starts its turn, keeping the nose ahead of the leader, pulling into position on the leader's INSIDE wing.

ALL AIRPLANES TAKE OFF IN THE ORDER OF JOINING FORMATION AT 30 SECOND INTERVALS. (TIMING FROM THE MOMENT PRECEDING AIRPLANE OPENS THROTTLE TO START TAKEOFF RUN)

Taxiing Out

At H hour, all ships start engines and stand by on interphone frequency. The formation leader checks with all planes in his formation. After this he calls the tower and clears his formation for taxi and takeoff instructions. As he taxies out No. 2 man follows, then No. 3, etc., each airplane taking the same place respectively on the ground that it is assigned in the air. As soon as the leader parks at an angle near the end of the takeoff strip, the other aircraft do likewise. At this point all aircraft run up engines and get ready for takeoff. The leader makes certain that everyone is ready to go before he pulls out on takeoff strip.

Takeoff

Formation takeoffs should be cleared from an airdrome in a rapid and efficient manner. Individual takeoffs will be made. Therefore, the following method is suggested.

The leader goes into takeoff position and takes off at H hour. No. 2 man starts pulling into position as soon as the leader starts rolling. When the leader's wheel leaves the runway, No. 2 starts taking off. (The time lapse is about 30 seconds.) The leader flies straight ahead at 150 mph, 300-500 feet per minute ascent, for one minute plus 30 seconds for each airplane in the formation. He levels off at 1000 feet above the terrain to prevent high rates of climb for succeeding aircraft. (Cruise at 150 mph.)

As soon as the leader has flown out his exact time, he makes a 180° half-needle-width turn to the left. The second airplane in formation assumes the outside or No. 2 position, while the third airplane assumes the inside or No. 3 position. The leader of the second element assumes position on the outside of the formation and his elements assemble on him in the same manner.

3-Airplane Vee

The 3-airplane Vee is the standard formation and the basic one from which other formations are developed. Variations of the Vee offer a concentration of firepower for defense under close control with sufficient maneuverability for all normal missions, and afford a bombing pattern which is most effective.

Flight of 6

A formation of 6 aircraft is known as a flight or squadron which is composed of two 3-airplane Vees. At least 50 feet vertical clearance will be maintained between elements in a flight and at least 50 feet horizontal clearance between the leader of the second element and wingmen of the first element.

From this basic squadron formation of 6 aircraft, the group, made up of 12 to 18 aircraft, is formed. Second or third flights will be echeloned right or left, up or down, with a vertical clearance of 150 feet and a horizontal clearance of 100 feet.

The high squadron flies 150 feet above and 100 feet behind the lead squadron with its second element stacked down and echeloned to the outside of the formation.

The low squadron flies 150 feet below and 100 feet behind the lead squadron with its sec-

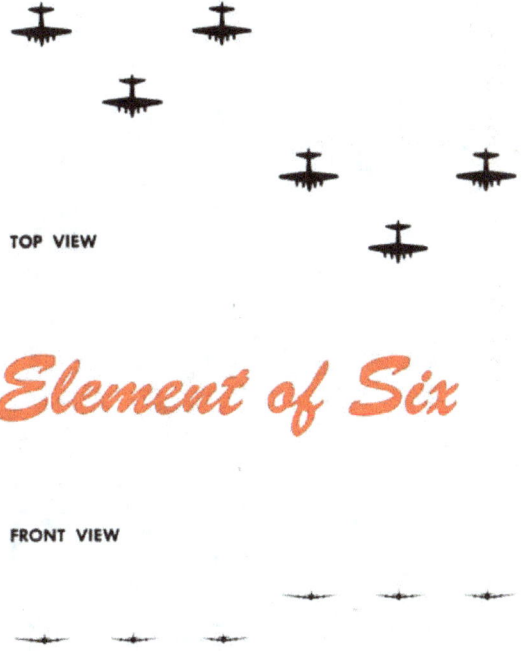

Element of Six

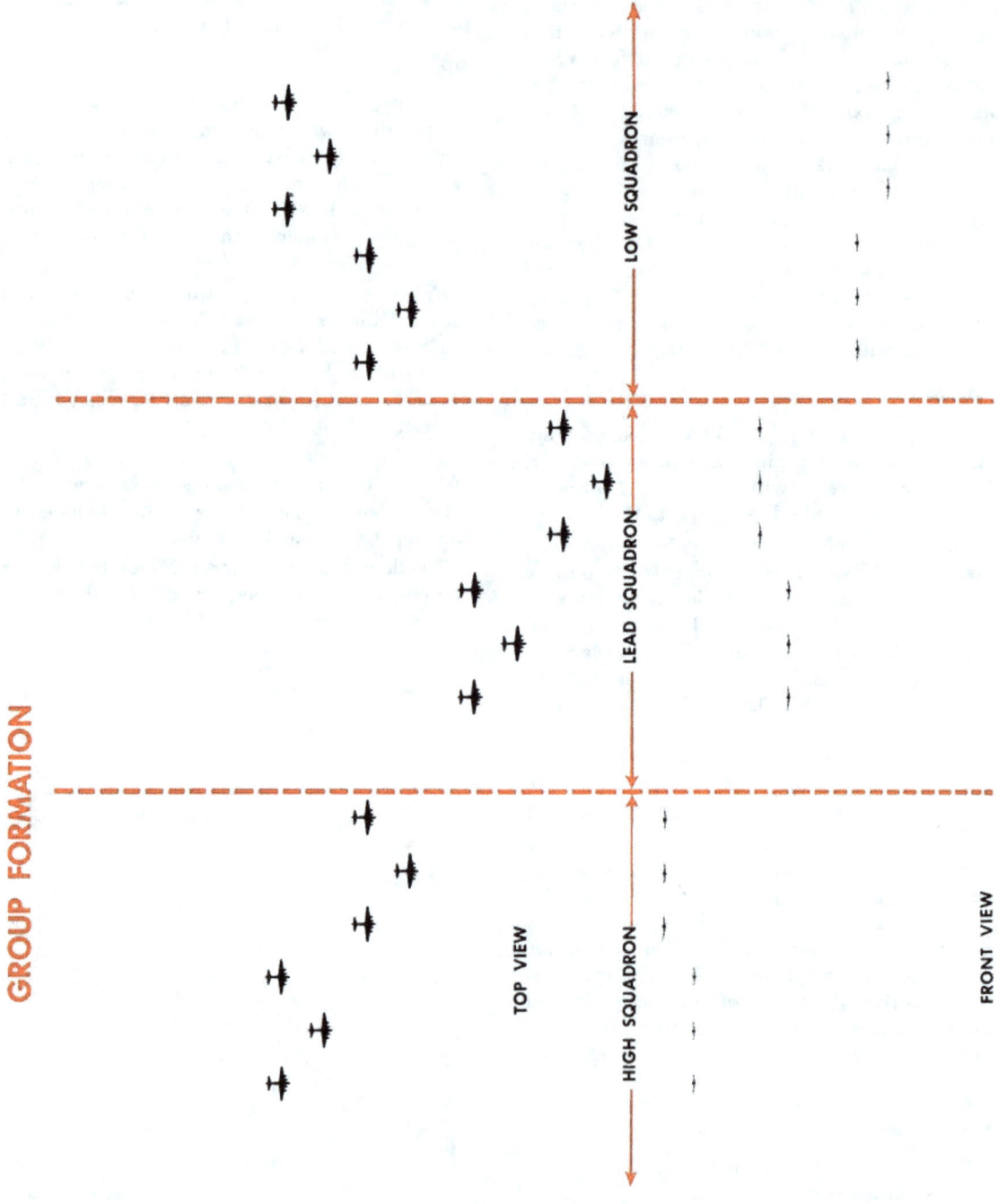

ond element stacked down and echeloned to the outside of the formation.

Flights may be placed in the high or low positions, as desired by the leader, by order over radio and receipt of acknowledgment. The flights simply go up or down in their respective positions. In this formation the positions of individual airplanes in each element will be those always flown in the 3-airplane Vee.

With but small variations, this basic formation can be changed to the combat formations used overseas. It is the job of training to teach a basic formation which can be readily understood and flown by students and easily adapted to tactical use.

Spacing of Wing Positions

It is particularly important for the leader to avoid violent maneuvers or improper positions which will cause undue difficulty for the wingmen.

The spacing of the wing positions in Vee formation is:

1. Vertically: On the level of the lead airplane.
2. Laterally: Far enough to the side to insure 50 feet clearance between the wingtips of the lead airplane and the wing airplane.
3. Longitudinally: Far enough to the rear to insure 50 feet clearance between the tail of the lead airplane and the nose of the wing airplane.

Turns in Vee formation will maintain the relative position of all airplanes in the element. In other words, the wing airplanes will keep their wings parallel to the wings of the lead airplane and on the same plane.

Trail

A formation is in Trail when all airplanes are in the same line and slightly below the airplane ahead. The distance between airplanes will be such that the nose of each succeeding airplane is slightly to the rear of the tail of the airplane ahead. If this distance is too great the propeller wash of the airplane ahead will cause difficulty in maintaining position. This formation will be used only when there are from 3 to 6 aircraft involved for changing the lead, for changing wingmen, and for peel-off for landing (optional).

Changing Wing Position

When changing from Vee to Trail, the wingman into whom a turn is made while in Vee assumes the No. 2 position in Trail, while the outside man is in the No. 3 position in the Trail. When returning from Trail to Vee, the No. 3 man in Trail assumes the inside position of the Vee. Remember this, for it is the procedure for changing from Vee to Trail and from Trail to Vee. Also, it provides a method for changing wing positions in a Vee formation.

It is often desirable for a leader to change the wing position of his formation, i.e., to reverse the right and left positions. If this maneuver is not executed properly in accordance with a pre-arranged plan, there is danger of collision. A safe plan is for the leader to announce on the radio that the formation will go into Trail on his first turn. If the turn is executed to the right, it will result in the inside man, or No. 2 wingman, being No. 2 in the Trail, and the outside man, or No. 3 wingman, being No. 3 in the Trail when the turn is completed. The leader will then announce that the formation will re-form in Vee when the Trail executes a turn to the right. This second turn to the right will re-form the Vee with wingmen reversed.

As stated above, this will result in the No. 2 man of the Trail assuming the outside position of the Vee, and the No. 3 man of the Trail assuming the inside position of the Vee. It is desirable for the leader to designate the ultimate position each wingman will assume prior to each turn in order to insure complete understanding.

Changing Lead

Formation will go into Trail from the usual 90° turn to the right or left. The leader of the formation will make a 45° turn to the left and fly that heading for approximately 20 seconds or until such time as a turn back will place him in the rear of the formation. When the

VEE-TRAIL-VEE

NO CHANGE IN WING POSITION

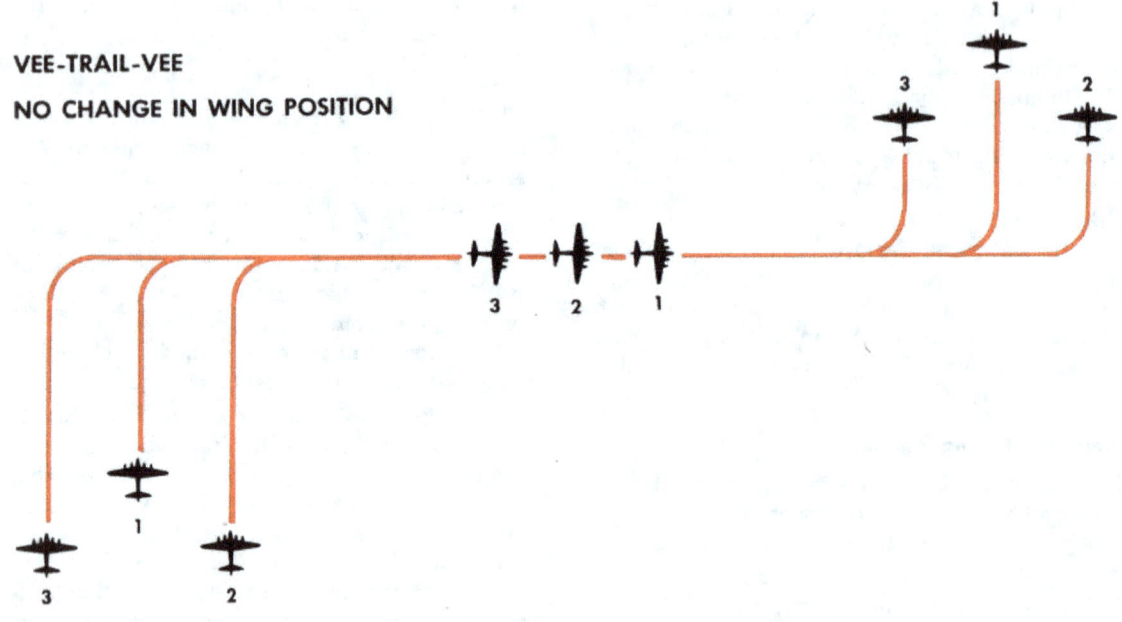

VEE-TRAIL-VEE

CHANGE WING POSITION

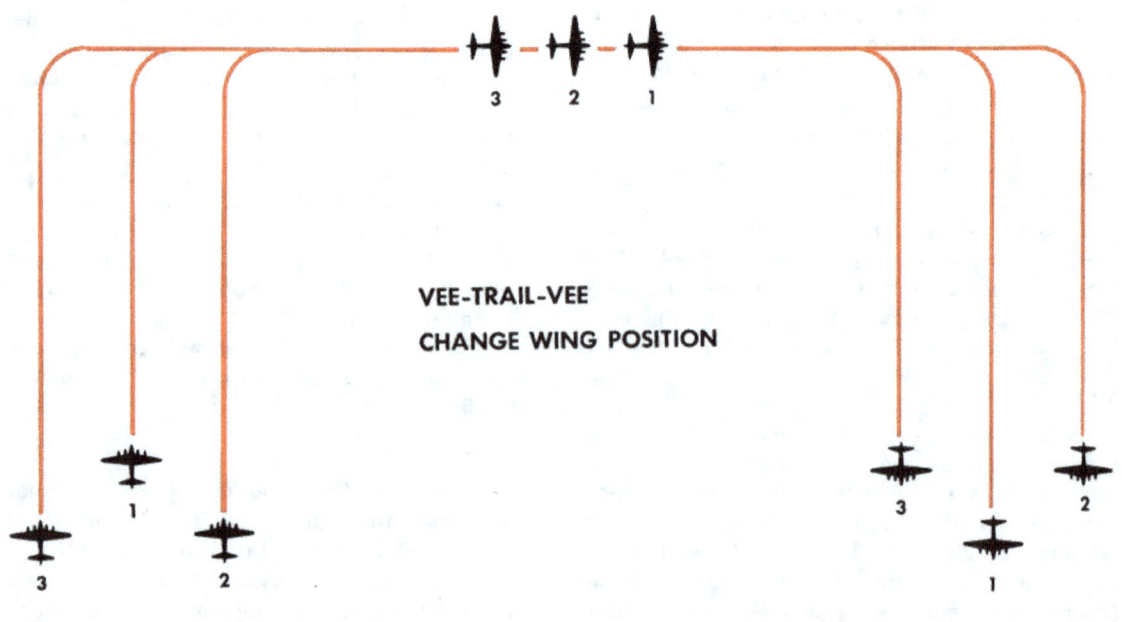

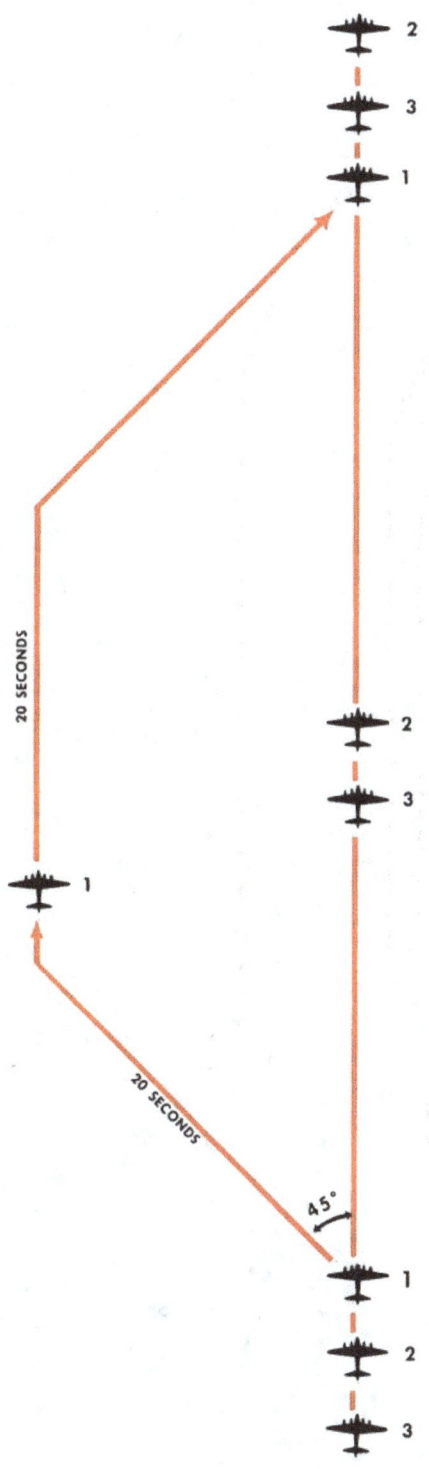

No. 1 airplane starts his 45° turn, the No. 2 plane in the Trail immediately becomes the leader of the formation and continues to fly straight ahead. At the end of 20 seconds, or thereabouts, the original leader turns back and takes up the No. 3 position in his element, or No. 6 position if in a flight of 6, and notifies the new leader that the maneuver is complete.

Landing

The formation will approach the field at an altitude of 1500 feet above the terrain in Vee in such a direction that two 90° turns either right or left can be made to bring the formation heading upwind in line with the runway on which the landing is to be made. The formation will go into Trail, stepped down, on the first 90° turn and the leader will order gears down as soon as the Trail has been formed, at which time the checklist may be started. The leader will then fly up to the runway and peel off to the left when he is directly over the spot on which he intends to land. Each succeeding plane will peel off without interval spacing achieved on first turn. The leader will put down ⅓ flaps, retard throttles, and make a continuous power let-down with just enough base leg to enable him to make a straight-away approach rather than a landing out of a turn, other ships in the formation spacing themselves and accomplishing the same approximate pattern of let-down and approach as their leader. There will be no more than 3 ships on the runway at the same time (one turning off, one midway, and one just landing).

Landing from Vee

The formation will approach the airdrome at an altitude of 1500 feet above the terrain into the wind up the landing runway, at which time the wheels will be ordered down by the leader and checklist accomplished. The second element will maintain assigned position echeloned to the right. The leader will call No. 3, when over the edge of the landing runway, to peel off, No. 3 acknowledging by peeling off. No. 1 follows; No. 2 following No. 1; No. 6 following No. 2 and so on. Approach and landing accomplished as outlined.

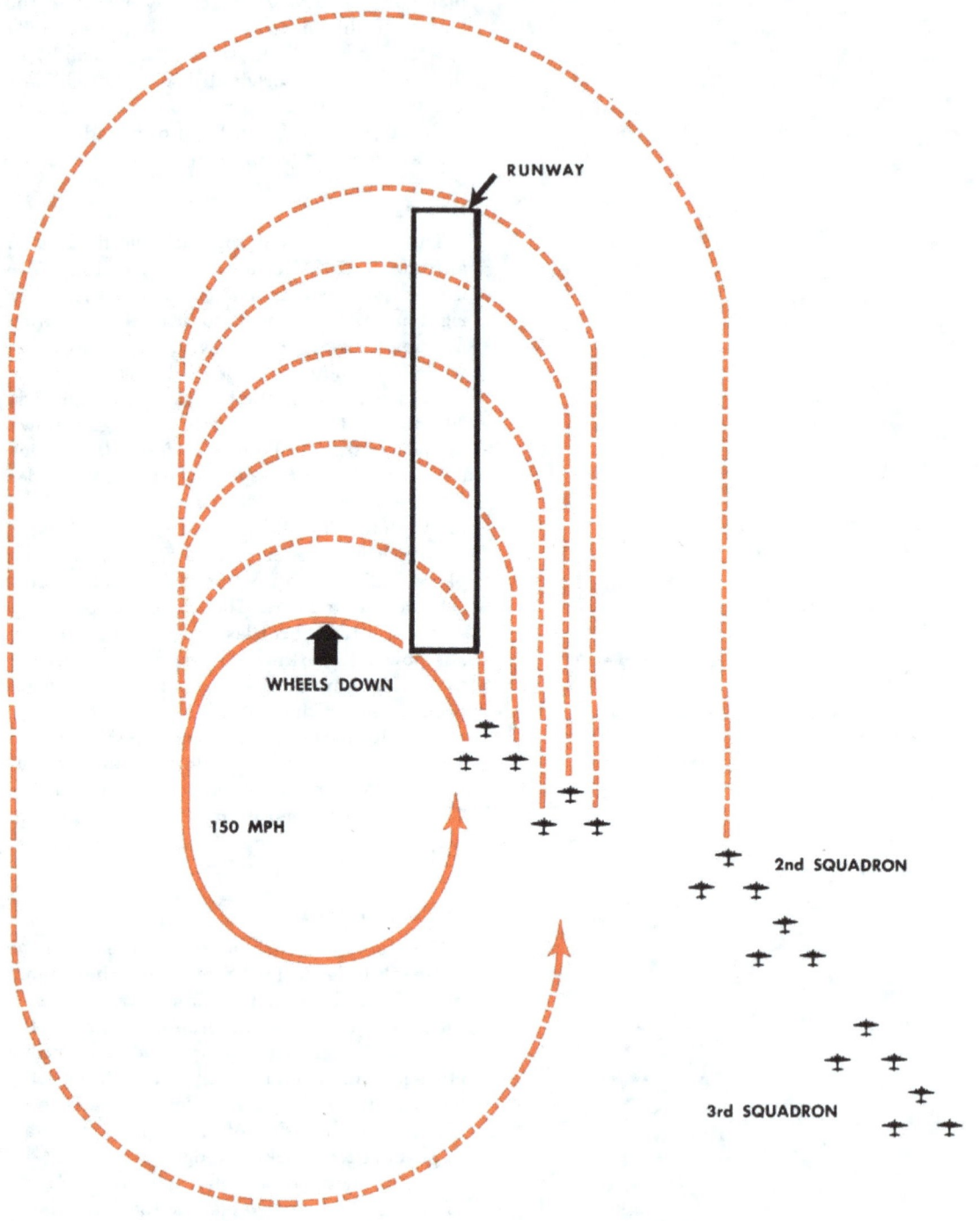

A Group Landing from Vee

The group will approach the airfield in an echelon of flights to the right. This echelon of flights will be accomplished by order of the leader by radio and acknowledged by the leader of flight indicated. The leader will have the formation with high squadron (flight) in second position, low squadron in third position still stacked down in low position, relative to leader's flight, but maintaining position on high squadron. Each flight will land individually, the lead flight landing first as previously outlined. The high and low flights will complete a 360° turn and land in turn as shown by diagram.

Conclusion

In conclusion, it should be stated that a good formation is a safe formation. An air collision is the result of carelessness or lack of clear understanding between members of the formation. If the simple rules, as outlined, are followed explicitly, there is no excuse for mistakes in the air. A mistake in formation flying may result in costly, irreparable loss of lives and equipment.

It should be reiterated that it is not a display of skill to fly too close; it is a display of bad judgment and lack of common sense.

TIPS ON FORMATION FLYING

1. Set rpm to minimum allowable for the maximum manifold pressure you expect to use.

2. At altitudes where superchargers are needed, set superchargers to give about 5″ more manifold pressure than the average being used.

3. Use throttles to increase and decrease power to maintain position. But when far out of position, or when catching up with a formation, increase rpm to maintain proper manifold pressure and rpm relationships.

4. When under attack, use all available power required to stay in formation.

5. In cross-over turns, keep a sharp watch out for your side of the airplane and have the copilot do the same on his side. The pilot or copilot (whichever can see the airplane below) should automatically take over the controls. If neither pilot nor copilot can see airplane below, then bombardier should give instructions by interphone.

6. In changing leads in practice formations or in Trail positions, avoid closing to proper formation position too rapidly. This can be dangerous.

7. In moving about in position, move the airplane in a direction that will not interfere with or endanger any other aircraft in the formation. In route formation, aircraft should be spread in width rather than depth in order to resume tight formation quickly.

8. At high altitudes, remember that rate of closure will be much more rapid than at low altitudes. It may be difficult to slow down quickly enough. Therefore, you will have to begin stopping the closure much sooner. On the other hand, acceleration is slower, so anticipation of change in position must be more acute.

9. Learn to anticipate changes in position so that only slight corrections need be made. Large corrections and constant fighting of the controls quickly wear out even a strong pilot.

10. Trim the ship properly. An improperly trimmed ship is difficult to hold in position.

11. Do not lock inboards and use outboards to maintain position. Use all 4 engines.

12. Whenever possible enter formation **from below** or on the level with the formation, **never from above**.

RESTRICTED

EMERGENCY PROCEDURES

FIRES IN FLIGHT

No emergency in an airplane is more serious than fire. Combat crews must always be conscious of the hazards involved in fire. They must be constantly on the alert for possible fire while in flight. They must be thoroughly familiar with methods of fire prevention and fire extinguishing.

Fires in flight can be prevented by more thorough preflight checks. Although most fires usually develop internally, many are caused by defects that could have been detected by visual inspection while on the ground. When making your visual inspection, look carefully for cracked or split exhaust stacks, excessive oil leakage, leaky primers, and gasoline fumes in the bomb bay or cockpit. All these are possible causes of fire in flight.

Be strict in forbidding smoking by crew members while transferring fuel in flight, and particularly when any gasoline fumes are detectable in the airplane.

Be careful in your checking procedure to see that the proper number of extinguishers are on board, and that the seals are not broken.

General Precautions

In case of fire during flight:

1. Warn all crew members to have parachutes attached in readiness for possible emergency use, and to stand by for orders.
2. If flying low, climb to safe altitude for possible bailout.
3. Determine whether airplane can be landed, or make plans for bailout.

Fire Inside the Airplane

1. Close all windows and ventilators.
2. If an electrical fire, cut electrical power to affected part.
3. If fuel line is leaking, cut fuel flow to affected line.
4. Make immediate use of either carbon dioxide or carbon tetrachloride extinguisher—preferably carbon dioxide, if available.
5. If necessary to use carbon tetrachloride, stand as far as possible from the fire. The effective range of this extinguisher is 20-30 feet. Remember that carbon tetrachloride produces a poisonous gas—phosgene. Do not use in a confined area, and do not stand near the fire when using it. A very small concentration of phosgene may prove fatal. After extinguishing a fire with carbon tetrachloride, open windows and ventilators.

Engine Fire in Flight

1. Alert the crew.
2. Cut fuel at tank by use of fuel cut-off switch.
3. Place propeller control in "HIGH RPM."
4. Apply full throttle to quickly scavenge engine and line of gasoline. With high rpm fuel pressure will drop almost before the pilot's hand can travel from the throttle to the feathering button. But if fuel pressure fails to drop (i.e., if the fuel shut-off valve has failed since the preflight), don't wait for a drop in fuel pressure.
5. Feather to cut the oil pressure.
6. Cut the generator and pull the voltage regulator to eliminate possibility of aggravating the fire if it happens to be an electrical fire.
7. Set selector and pull CO_2 charges (if installed).
8. Complete after-feathering procedure (see p. 143).

If it is a gasoline fire, cut off the source of fuel by using the fuel shut-off switch.

If it is an oil fire, cut the source of fuel by feathering.

If it is an electrical fire, remove the cause by cutting generator and pulling voltage regulator.

Engine Fire on the Ground

1. Close fuel shut-off switch.
2. Place propeller control in "HIGH RPM."
3. Apply full throttle.
4. Feather the engine.
5. When propeller stops turning, cut off master switch.

Be sure that cowl flaps are open so that the fire guard can effectively use external extinguisher.

If necessary, set and pull the engine fire extinguisher.

The next move is to get out of the airplane.

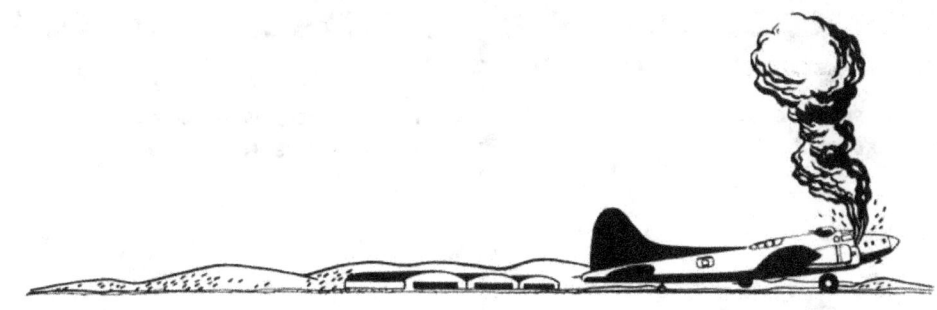

RESTRICTED

MAXIMUM PERFORMANCE TAKEOFF

The purpose of this maneuver is to take off in a minimum distance—in other words, to make a short-field takeoff.

1. Line up with the runway and complete checks.

2. Put down ⅓ flaps.

3. Hold brakes, raise elevators, and increase throttles to 35" manifold pressure.

4. Release brakes and increase power by steadily and continuously opening throttles.

5. Hold airplane in 3-point position during entire takeoff run.

6. Keep cowl flaps from ½ to fully open for takeoff even in coldest weather.

7. When airspeed is sufficient, ease the airplane into the air by pulling back slowly and steadily on the control column. If the airplane is properly trimmed, takeoff will require little back pressure.

8. When airborne, leave the flaps in the ⅓ down position until all obstacles have been cleared and you have attained 130 mph IAS.

Directional Control

During the earliest stage of the takeoff run, the airplane is inherently stable. **It will tend to move straight ahead in the direction it was pointing when brakes were released.** For this reason it is extremely important to line up properly before attempting the takeoff.

Do not use brakes to maintain directional control. Use rudder and throttle if necessary, as in a normal takeoff. Rudder remains relatively ineffective until considerable speed is attained. The best procedure is to establish the proper direction by lining up properly before takeoff.

3-Point Takeoff

Two general warnings concerning the 3-point takeoff must be mentioned to new pilots.

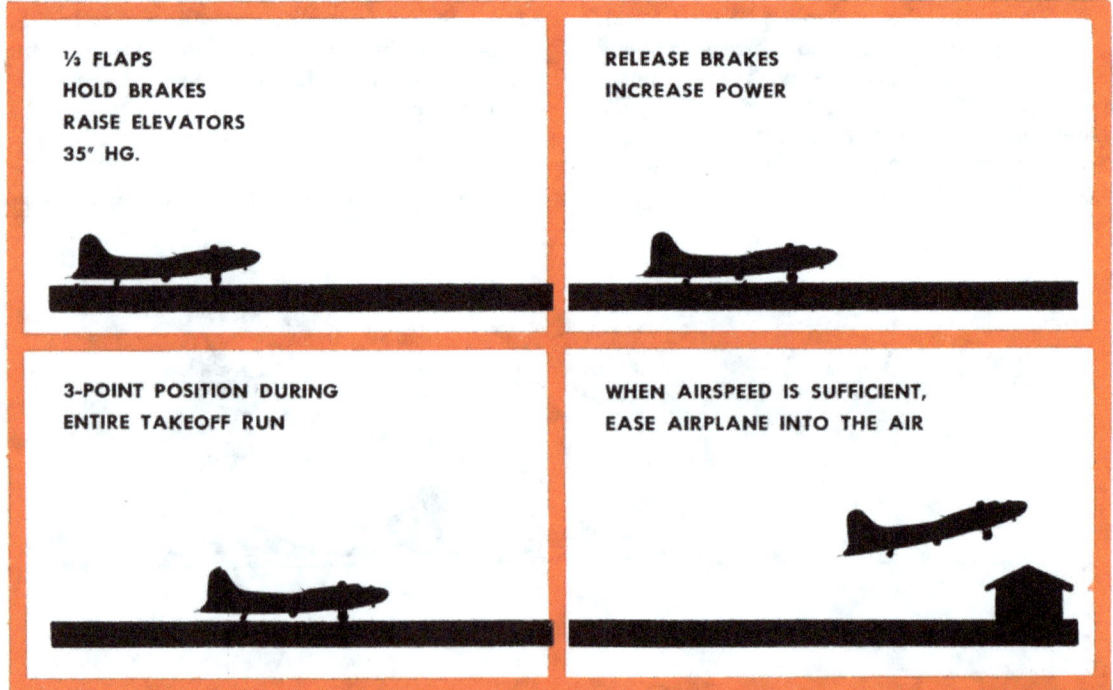

⅓ FLAPS
HOLD BRAKES
RAISE ELEVATORS
35" HG.

RELEASE BRAKES
INCREASE POWER

3-POINT POSITION DURING
ENTIRE TAKEOFF RUN

WHEN AIRSPEED IS SUFFICIENT,
EASE AIRPLANE INTO THE AIR

RESTRICTED

First, the airplane can be nosed over by holding brakes and applying high power **unless the tail is held down by the elevators.**

Second, never allow the 3-point takeoff to become a one-point takeoff. Be sure you know the feel of the airplane in a 3-point attitude. Otherwise, you may hold the tail down too far and too long, thereby causing the airplane to stall off the ground **tail last.**

Use of Flaps

Putting down ⅓ flaps before releasing brakes (rather than waiting until 70 mph airspeed has been attained) is recommended for 2 reasons:

1. Takeoff is the most critical stage of flight operations. Waiting until you are ⅔ of the way down the runway to lower ⅓ flaps only complicates the procedure, and diverts attention from the actual takeoff to a dangerous degree.

2. Experience shows that even on a concrete runway there is actual improvement in takeoff performance by carrying as much weight as possible by winglift (i.e., with the use of flaps) instead of on the wheels. The advantage is even more marked where the takeoff surface is rough or soft.

Other Aids to Maximum Performance Takeoff

1. Lighten the airplane, retaining only enough fuel to reach next landing place. Throw overboard tools, guns, miscellaneous equipment.
2. Set propeller governors for 2760 rpm.
3. Set turbos for 55″ manifold pressure.
4. Take off with as low a cylinder-head temperature as possible to avoid detonation.
5. Climb at the same speed as the airplane leaves the ground.

MAXIMUM PERFORMANCE LANDING

Here the purpose is to land the B-17 in the shortest possible distance.

1. Bring in the airplane for a normal 3-point landing.
2. Open cowl flaps on approach. On contact with runway, have copilot retract wing flaps.
3. Raise the tail with elevators to reduce angle of attack and thus hold more weight on the main wheels.
4. Apply brakes gradually but firmly until you have applied maximum pressure possible without skidding the tires. Remember that jamming on the brakes may cause the airplane to nose over.
5. Keep the tail off the ground as long as possible.

Be sure that brakes are not applied before the weight of the airplane has settled on the

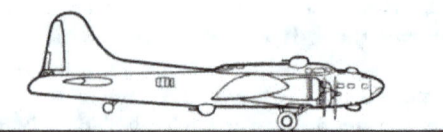

KEEP TAIL OFF RUNWAY AS LONG AS POSSIBLE

runway. Resultant skidding will blow out the tires almost immediately.

If the airplane has to be groundlooped at the end of the runway, unlock the tailwheel **while there is no side load on it.** Any side load on the tailwheel at this time will bind the locking pin.

Clearing Obstructions

When an obstacle must be cleared in order to make a maximum performance landing on a short field:

1. Clear the obstacle at minimum safe airspeed.

2. Immediately after clearing the obstacle, steepen the glide so that it can be broken as soon as possible and contact made with the ground.

3. Immediately after contact, bring the airplane to the tail-high attitude and apply maximum braking power as described above. The slightly increased landing speed will be more than offset by the gain of additional runway space and over which brakes can be used.

NO-FLAP LANDING

If the flaps cannot be lowered for landing, you can make a no-flap landing safely in the B-17.

1. Fly the traffic pattern just as you would for a normal approach with full flaps, **but maintain minimum airspeed of 140 mph until you are on the final approach.**

2. Set a power glide with manifold pressure of approximately 15″ Hg., and an airspeed of 125 mph.

3. The airplane will land at between 105 and 115 mph, depending on the gross weight. Therefore, be careful not to allow airspeed to drop below 120 mph until after breaking the glide.

4. Start the power glide at a point approximately ½ mile farther from the field than you would normally for a full-flap landing.

5. This landing is extremely hard on brakes. Difficulty may be encountered in stopping the airplane before you run out of runway. Start using brakes immediately after the airplane has settled on runway. Several applications may be necessary.

LANDING WITHOUT FLAPS

MINIMUM AIRSPEED 140 MPH

POWER GLIDE
MANIFOLD PRESSURE 15″ HG.
AIRSPEED 125 MPH

LANDING SPEED BETWEEN 105 AND 115 MPH

EMERGENCY OPERATION

OF LANDING GEAR, WING FLAPS, AND BOMB BAY DOORS

If you cannot operate landing gear, flaps, or bomb bay doors by the usual electrical means:

1. Place switches in neutral and check the landing gear and flap circuit fuses.
2. Cranks for manual operation are stowed on the aft bulkhead of the radio compartment. Extensions for use in operating engine starters, bomb bay doors, and wing flaps are stowed adjacent to the cranks.

Operation of Landing Gear

You can operate each main landing gear separately through the hand crank connections in the bomb bay. One connection is to the left of the door in the forward bulkhead, the other is on the right.

To raise the wheel, insert the hand crank into the connection. Direction of rotation will vary with type of retracting motor installed.

Be sure the landing gear electric switch is in the "OFF" position before attempting to raise or lower wheel by hand cranking.

Emergency Operation of Tailwheel

Use the same crank for manual operation of the tailwheel. Insert the crank into the connection in the tailwheel compartment and rotate as desired.

Be sure the landing gear electric switch is in the "OFF" position before attempting hand cranking.

Emergency Operation of Wing Flaps

Lift the camera pit door (in the floor of the radio compartment) and insert the crank into the torque connection at the forward end of the pit. Rotate the crank clockwise to lower flaps, counter-clockwise to retract them.

Be sure the electric switch is "OFF" before hand cranking.

Emergency Operation of Bomb Bay Doors

Insert the hand crank into the connection in the step at the forward end of the catwalk in the bomb bay. Rotate clockwise to close the doors, counter-clockwise to open.

Emergency Bomb Release

An emergency bomb release handle is at the pilot's left, and there is another at the forward end of the catwalk in the bomb bay.

Pull the handle through its full travel. The first part of the stroke unlocks the bomb bay doors independently of the retracting screw and permits them to be held open by wind pressure. The second half of the stroke releases all external and internal bombs, in salvo and unarmed.

To retract the doors after emergency release of bombs (see illustration, p. 134):

1. If the spring in the emergency release mechanism (under the hinged door beneath the pilot's compartment floor) has not retrieved the linkage entirely, re-set by pushing the hinge of the link.
2. Operate the retracting screws electrically (or manually, if necessary) to the fully extended position. This will engage the latches between the screws and the door fittings.
3. Now close the doors in the normal manner.

RESTRICTED

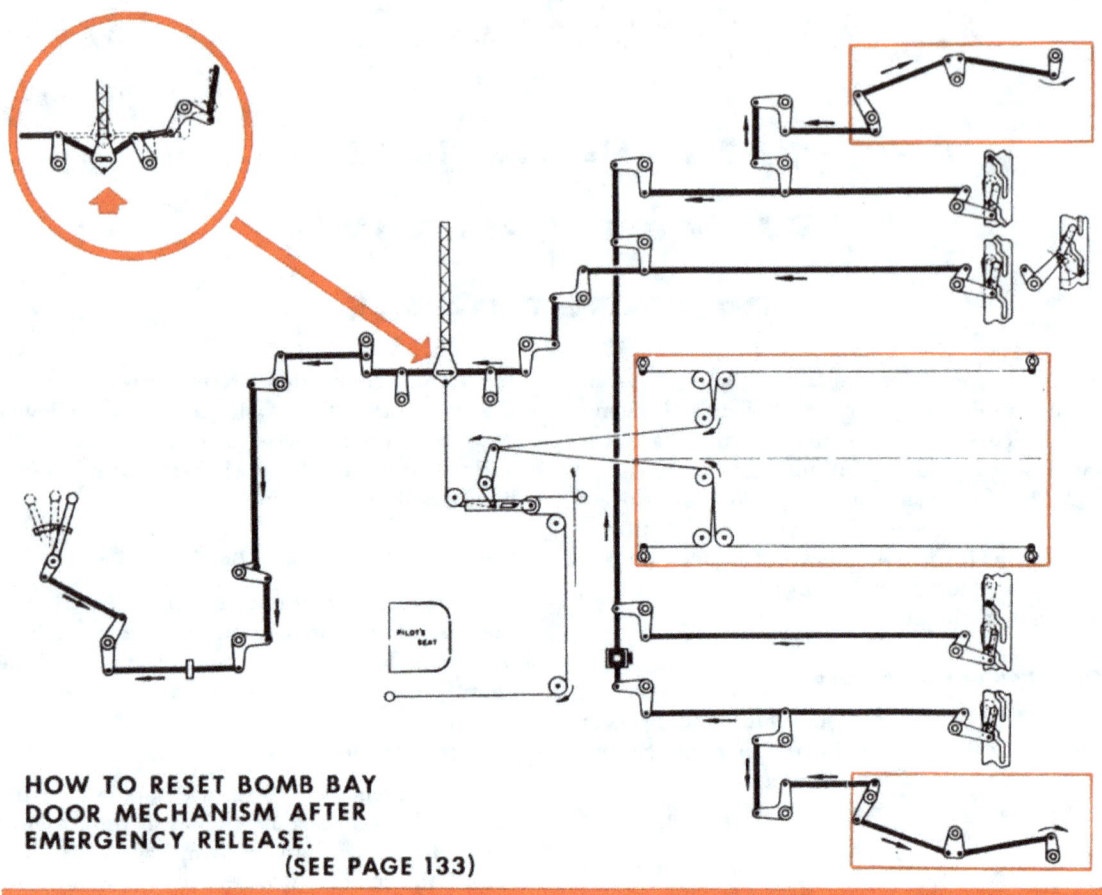

HOW TO RESET BOMB BAY DOOR MECHANISM AFTER EMERGENCY RELEASE.
(SEE PAGE 133)

DROPPING THE BALL TURRET IN FLIGHT

When preparing to bring the B-17 in for an emergency wheels-up landing, it is desirable to drop the ball turret in order to minimize damage to the fuselage when it hits the ground.

It is both safer and easier to release only the turret ball itself, leaving the supporting yoke intact. Only 2 tools—a crescent wrench and a hammer—are needed to do the job. Two men can accomplish it in approximately 20 minutes.

1. Point the guns aft or down and remove the azimuth case, which is held by 4 bolts.

2. Remove the safety retaining hooks. These 4 hooks can be broken off with a hammer, or they can be removed with a socket wrench if one is readily available.

3. Remove the 12 yoke connection nuts. The turret may hang momentarily on the fire cut-off cam, but a firm kick on the aft side of the ball will dislodge it.

It is desirable, but not absolutely necessary, to disconnect the electrical plug and oxygen line before removing the yoke nuts.

RESTRICTED

If time permits, the computing sight can be salvaged by first entering the turret and disconnecting the 3 flexible drive cables at the left, right, and far side of the sight. Free the sight by removing the right retaining rod and disconnecting the electrical plug. Removal of the sight may add approximately 20 minutes to the time, making the total time necessary for the operation about 40 minutes.

Remember these 2 rules for making emergency landings:

1. When landing the B-17 **with wheels retracted,** drop the ball turret.

2. When belly-landing a B-17 in which a **chin turret** is installed, **retract the tailwheel also.**

FOR MINIMUM STRUCTURAL DAMAGE, MAKE BELLY LANDINGS:

    a. **WITHOUT BALL TURRET**

    b. **WITH TAIL WHEEL DOWN**

    c. **WITH ¾ FLAPS DOWN**

STUDY DETAILED PROCEDURES ON PAGES 134-137

LANDING DISABLED AIRCRAFT

Landing on One Flat Tire

1. Bring the airplane in at a normal glide, using full flaps, for a normal landing.
2. Don't make an effort to land on the good tire. However, if one wheel is to come in contact with the ground first, it should be the good tire.
3. Hold elevators all the way back immediately after landing.
4. Use the brakes on the good tire only.

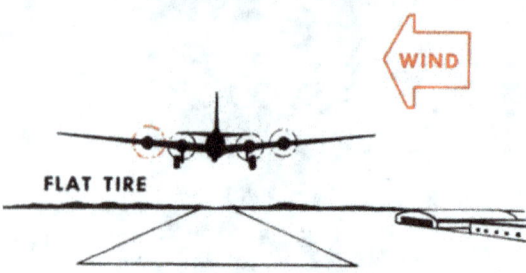
FLAT TIRE / WIND

5. Use the outboard engine **on the side of the flat tire** to counteract any tendency to ground-loop on that side.

If possible, land the airplane crosswind, with the wind coming from the side with the good tire. This crosswind tends to make the airplane turn into the wind; but the effects of the wind and the flat tire tend to equalize, and less difficulty is experienced in keeping the airplane straight.

Landing with Cracked or Wobbling Wheel

1. Land directly into the wind, making a normal 3-point landing.
2. Use brakes on the good wheel only.

Experience shows that in most cases the damaged wheel will stand up for a final landing, with no damage to the airplane.

Landing with Bent Drag Link

1. Make a normal approach, with full flaps, and with the good wheel extended.
2. Land at the very beginning of the runway to allow all possible room for the landing roll.

3. Don't use brakes. Use throttles to keep the airplane straight until they are no longer needed or until the gear collapses.

In most emergency landings of this type (where the above procedure has been used) the airplane has slowed down or stopped completely by the time the gear collapsed. Damage to the airplane itself has been slight: a bent outboard wing panel, or damaged propellers on one side. The ball turret has seldom suffered.

Landing With Broken Drag Link
(B-17 With Ball Turret Installed)

For an emergency landing under these conditions, complete these preliminaries:

1. Tie down all loose equipment.
2. Open waist gunner's window.
3. Open radio compartment hatch.
4. Open pilot's and copilot's windows.

These precautions must be taken to facilitate quick exit in case of fire.

Use up or dispose of unneeded fuel. Then follow this procedure:

1. Leave the good wheel down. Leave the tailwheel fully extended. Allow the damaged gear to hang free.
2. Bring the airplane in on a normal power approach, directly into the wind, with speed slightly above normal.
3. Put down only ¾ flaps. It has been found that with flaps retracted only ¼ they sustain no damage at all.
4. Make what would be a normal 3-point landing, except for the fact that one wheel is dangling free.

Note: Use the above procedure also in the B-17G (ball turret and chin turret installed) when landing with a broken drag link.

Landing With Broken Drag Link
(B-17 With Ball Turret Removed)

In a B-17 from which the ball turret has been removed, a landing with broken drag link can be made in the following manner:

1. Complete precautionary measures for crash landing as outlined on p. 156.

2. Retract the good gear. Leave the tailwheel extended. Allow the damaged gear to hang free.

3. Bring the airplane in on a normal power approach, directly into the wind, with speed slightly above normal.

4. Put down ¾ flaps, since it has been found that with flaps retracted ¼ they sustain no damage at all.

5. Make what would be a normal 3-point landing except that one wheel is dangling free.

Immediately after contact with the ground, pull back on the control column sharply. The result will be a slight lifting of the nose.

Upon impact with the ground, the wheel on the damaged gear begins to spin, the main strut of the gear hits the stop and bounces forward. If you have timed your back pressure on the control column properly, the wheel will come forward of the vertical position. When the weight of the airplane settles, the wheel rocks forward and into its nacelle. The airplane then slides down the runway on its belly.

As the landing is being made, either pilot or copilot watches the position of the damaged wheel. If the maneuver fails to force the wheel forward and into its nacelle, it is possible to advance the throttles and go around again.

RUNAWAY PROPELLERS

The most important fact to keep in mind about a runaway propeller is **not to feather it** until you have tried the 2 procedures which should give you control of it. Drill your copilot in these procedures, so that he will understand his part in controlling a runaway propeller.

What Causes a Propeller to Run Away

When a propeller runs away, it simply means that the propeller governor has failed to hold the propeller at its constant rpm setting. Thus, before takeoff, when engines are idling, the propeller is in "HIGH RPM." Sudden application of power may cause a propeller to exceed the governor limit speed before the governor has a chance to increase pitch. The governor cannot regain control until you throttle back and give it a chance. This is usually the case with a runaway propeller.

However, if you have complete governor failure, you may not be able to regain control with throttles or with propeller controls, and will have to use the feathering button intermittently as described in the following procedures.

Preventive Action

The best way to cope with a runaway propeller is not to get one! Carefully observe tachometer reaction during run-up. Don't jam on power during takeoff. Apply it smoothly.

How to Regain Control

Always try this first, during takeoff and in flight. It may give you immediate control over

**WITH BALL TURRET
LAND WITH GOOD WHEEL DOWN**

**WITHOUT BALL TURRET
MAKE A BELLY LANDING**

RESTRICTED

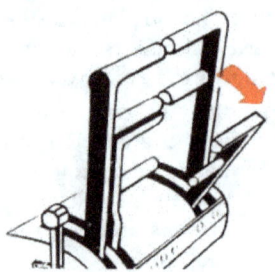

the runaway propeller so that you can obtain a normal rpm setting.

First procedure:

1. Reduce the throttle. This is the first step necessary to slow down the propeller.

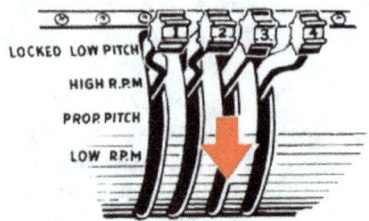

2. Push down propeller control to cut rpm.
3. If this works, re-set your throttle, keeping close watch on rpm. If it fails, resort to the second procedure.

Second procedure: (This procedure is recommended for takeoffs and for heavily loaded airplanes because it gets more power.)

1. Reduce the throttle.

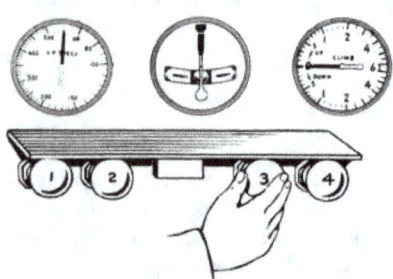

2. Copilot, at pilot's direction, pushes in the feathering button, holds it in, and watches rpm. (**Be sure to get the right feathering button,** or you'll be short 2 engines. Take your time!)

3. As propeller rpm decreases, increase the throttle to obtain climbing manifold pressure and 2500 rpm.

4. When rpm reaches 2500, forcibly pull the feathering button out. This will keep the rpm from decreasing further. If the governor does not take control of rpm, it will immediately start back up.

5. When propeller reaches 2760 rpm, push feathering button in, repeating procedure to keep rpm between 2500 and 2760 and maintain desired manifold pressure. Continue this until you attain an altitude where you can go around safely and land, or where you can feather the propeller.

Caution

Don't be in a hurry to feather. If either of these procedures is keeping the propeller below 2760 rpm, you are getting some power from the engine, possibly as much as 15% with the throttle retarded, and up to 65-70% by using the second procedure.

OVERSPEEDING TURBOS

1. Throttle back affected engine.
2. Close turbo control. (If electronic system is installed, change amplifier on turbo.)
3. Try to maintain desired power with throttle.
4. Try to re-set the turbo for operation without overspeeding. Usually, there is a position where the turbo will stay within operating limits.

During this operation, maintain directional control with rudder. Never throttle back the opposite engine unless full rudder fails to hold the airplane on a straight course.

On takeoff, never feather an engine if the turbo or propeller can be brought under control. Bear in mind that you will need all the power you can get.

BRAKE OPERATION WITH HYDRAULIC PUMP FAILURE

If pressure in the main hydraulic system and emergency system (if installed) is completely lost because of failure of the hydraulic pump or shot-up lines, use the hand hydraulic pump located to the right of the copilot on the floor.

1. Place the hydraulic selector valve in the "NORMAL" position. Set the star valve of the emergency system (if installed) in the "CLOSED" position.

2. When ready to apply brakes, **depress the pedals** and have the copilot operate the hand hydraulic pump. The pump will have no effect unless the pedals are depressed.

3. No resistance will be felt to the first few (3 to 10) strokes of the hand pump. The copilot must remember to keep on pumping because no braking action is possible until resistance develops.

Bear in mind that hand pump operation supplies direct action to the brakes. No pressure is being stored up in the accumulator. This procedure is unnecessary in later models of the B-17F equipped with the emergency brake system.

When you release the brake pedals, all pressure will be dissipated through the brake metering valves to the return lines. To re-apply brakes, the foregoing procedure (use of brake pedals, use of hand pump) must be repeated in its entirety.

EMERGENCY HYDRAULIC SYSTEM
INSTALLED ON B-17F ONLY

The emergency hydraulic system consists of an additional accumulator charged by the electrically driven pump, and 2 manually operated metering valves located in the roof of the pilot's compartment.

The system operates the brakes only: the left hand lever controls the left wheel brake, the right hand lever controls the right wheel brake. Pulling the handles downward directs pressure from the emergency accumulator through the auxiliary brake lines. This provides braking control in the event that the main hydraulic system has failed.

If it is necessary to service the system, follow this procedure:

1. Manual shut-off valve to "CLOSED."
2. Selective check valve to "NORMAL."
3. Determine pressure in the emergency accumulator.

Do not pump the emergency hydraulic brakes. Pressure in the emergency system is limited (approximately 4 applications) and pumping will result in early loss of emergency brake control.

If pressure in the emergency system is lost, use the hand pump (on the floor to the copilot's right). Make sure that the hydraulic selector valve is in the "NORMAL" position. With the valve in this position pressure is bypassed around relief valve directly to brakes.

Selector valve must always be in the "NORMAL" position for emergency operations.

FEATHERING PROPELLERS

Feathering mechanism is incorporated in propellers for two reasons: (1) to reduce drag when the airplane must continue flight with only 3 or 2 engines operating; (2) to eliminate vibration of a damaged engine that might otherwise weaken the airplane's structure.

To Feather or Not to Feather?

Feathering is an important and valuable procedure—when needed. When you're satisfied that feathering is indicated, and you're sure you know what you're doing, don't be afraid to feather the engine. But don't be too hasty in hitting that feathering button. Be sure you know when to feather. Be sure you clearly understand the advantages to be gained by feathering. **Be sure you feather the proper engine.**

When to Feather

When confronted with engine trouble, and the question of whether or not to feather the damaged engine, follow these rules.

1. Be calm, think clearly, move slowly. Your problem is to decide whether or not to feather; and, if feathering is indicated, to feather the proper engine.

2. Generally, an engine losing power should not be feathered so long as it is still producing power and is not vibrating excessively. If you are not sure the engine is still operating, engage the turbo. A rise in manifold pressure will indicate whether the engine is still putting out some power. In a possible emergency, don't throw away usable power.

3. If an engine is running rough, try a change of power setting. Try a change of mixture control position; also check intercooler control position. These checks will sometimes produce smoother operation.

4. Be sure that the real trouble lies in the engine, not in your engine instruments. If oil pressure drops, for instance, check your oil temperature gage: oil temperature will rise if anything is radically wrong (unless you're out of oil). If your oil pressure drops to about 30 lb., however, feather the propeller while you still have oil, and ask questions later.

5. Before deciding that you have a runaway propeller, set a definite rpm at which you will

feather (2760 rpm maximum). Unless rpm reaches that danger point, continue to operate the engine with reduced manifold pressure, especially on takeoff.

6. Once you have decided to feather, be sure that you feather the damaged engine and not one of your good engines by mistake. In addition to checking manifold pressure, rpm, oil pressure, and oil temperature, look for the other signs which indicate the location of the faulty engine.

7. Tendency to turn will indicate whether the faulty engine is on the left or right side.

8. Noticeable vibration often will identify the faulty engine.

If there is no reserve supply, and oil pressure falls to 30 lb., feather the engine at once. If a reserve supply is available, watch for a rise in oil temperature before feathering: this will indicate whether oil pressure is really low.

How to Feather in an Emergency

When you have decided to feather, and you're sure that you're feathering the proper engine, your immediate procedure is as follows:

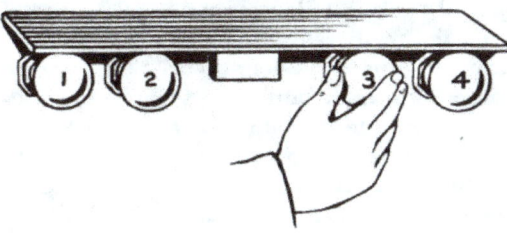

1. Close the propeller feathering switch.

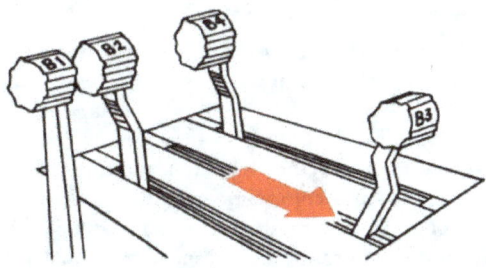

2. Turn turbo supercharger control "OFF."

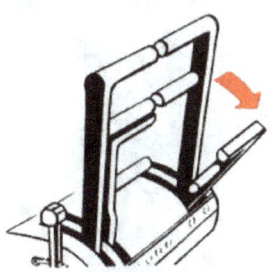

3. Close throttle.

4. Move mixture control to "IDLE CUT-OFF."

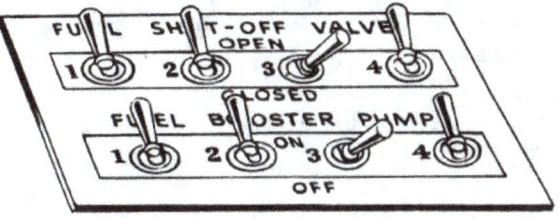

5. Switch fuel shut-off valve "CLOSED," booster pump "OFF."

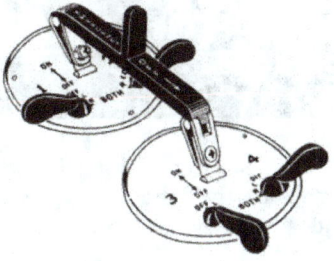

6. After propeller has stopped, turn ignition switch "OFF."

When these immediate steps have been taken, continue with this clean-up procedure.

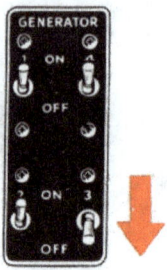

1. Turn generator "OFF."

2. If landing gear is down, retract it unless you can land immediately.

3. Have copilot adjust mixture controls on the other engines, and increase rpm as required. Increase manifold pressure.

4. Trim the airplane.

5. Change vacuum selector position, if necessary.

6. Close cowl flaps on the dead engine. Adjust cowl flaps on the other 3 engines to maintain cylinder-head temperatures within safe limit.

7. Transfer fuel from the dead-engine tank, if needed.

Emergency Measures

1. Tune the radio compass to nearest stations, so that you can use the radio compass needle for making turns and if instruments fail through loss of vacuum, you can maintain direction by homing from one station to another.

2. If vacuum is out and you have to fly on instruments, turn the automatic pilot "ON." Refer to tell-tale lights to maintain level flight attitude. **Don't turn on rudder, elevator or aileron switches.** This is used only as an additional aid. Otherwise use airspeed, ball, and compass.

Normally the feathering switch is released by hydraulic pressure built up in the system after the propeller has reached the full feathered position. Sometimes viscous oil in the propeller system builds up this trip-out pressure prematurely, preventing full feathering. If this happens, hold the feathering switch down until the propeller is fully feathered.

Accidental Unfeathering

In some cases hydromatic propellers have begun to unfeather almost immediately after reaching the full feathered position. This is because the switch failed to cut out automatically when the feathered position was reached.

Should this condition occur, pull out the feathering switch button as soon as the propellers begin to unfeather. Leave it out for 2 or 3 seconds, then close the switch again. When the full feathered position has been reached (indicated by the cessation of windmilling) pull the feathering switch button out again. This will prevent further unfeathering.

Failure of Feathering System

Total loss of engine oil in combat, or line failure in the engine oil system, will make feathering impossible (unless auxiliary supply is available). If normal feathering is impossible, try to make the propeller windmill at the lowest possible rpm. Since windmilling is proportional to airspeed, it can be reduced to a minimum by reducing airspeed to 20-30 mph above stalling speed (i.e., to approximately 120-130 mph IAS).

1. Place propeller control in "LOW RPM."
2. Place mixture control in "IDLE CUT-OFF."
3. Turn ignition switch to "OFF" position.
4. Set throttle to fully closed position.
5. Fuel shut-off switch: "OFF."

Vibration can be reduced or minimized by flying at the absolute minimum airspeed.

Engine Seizure

Frequently loss of oil for lubrication will cause the engine to seize and stop suddenly. In some cases of engine seizure the reduction gear housing will break, allowing the propeller, propeller shaft, and reduction gearing to fall off. In other cases, only the reduction gears will be stripped. This relieves the propeller of engine drag and permits it to windmill.

Emergency Unfeathering

Never unfeather a propeller of a faulty engine unless it is needed for landing or continued flight. If the propeller was feathered because of engine damage, remember that unfeathering may result in still further damage.

Be especially careful in starting and warming up a cold engine. Oil drains into the bottom cylinder of a dead engine, and structural damage may result from re-starting the engine.

When practicing feathering, don't allow the propeller to remain in the feathered position for more than 5 minutes. Under cold weather conditions, unfeather the propeller at once.

How to Unfeather

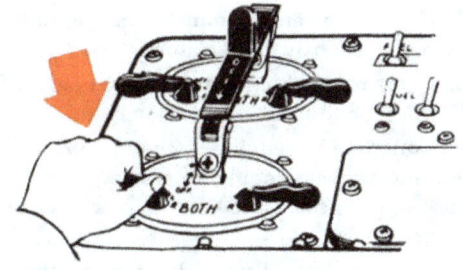

1. With throttle closed, turn ignition switch "ON." (Except in B-17G.)

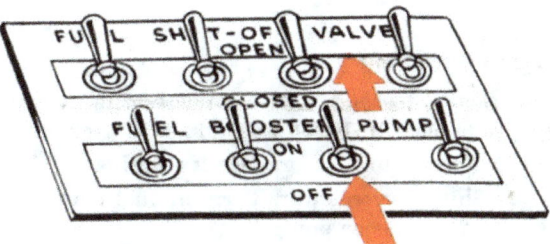

2. Switch fuel shut-off valve "OPEN," booster pump "ON."

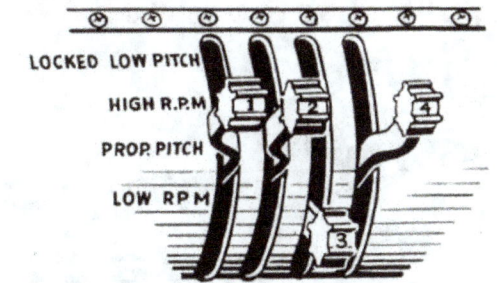

3. Set propeller control to "LOW RPM."

4. Close feathering control switch, and keep it closed until tachometer reads 800 rpm. Then pull out propeller control switch.

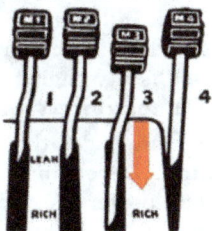

5. Place mixture control in "AUTO-RICH" position.

6. Allow engine to operate at 800 rpm, until 100° cylinder-head temperature is obtained. Then operate throttle gradually until engine speeds up to minimum rpm, or speed at which governor is set.

7. Adjust mixture, rpm and throttle to desired settings, and synchronize propellers.

Practice Feathering

Practice feathering and unfeathering at an altitude 5000 to 10,000 feet above the terrain.

The procedure for **practice feathering**:

1. Shut off generator. Turn off fuel shut-off valve and booster pump.
2. Close the throttle.
3. Close supercharger.
4. Mixture control to "IDLE CUT-OFF."
5. Close propeller feathering switch.
6. Turn ignition switch "OFF" after propeller stops turning.
7. Complete clean-up procedure.

ONE-ENGINE FAILURE ON TAKEOFF

The failure of one engine on takeoff will not present much difficulty if intelligent action and proper technique are applied immediately.

If the failure occurs just after takeoff, use what power is left in the engine to attain critical 3-engine speed. Think calmly, act positively to determine the faulty engine, and then feather. **Be sure it is the correct engine.**

At the same time, get the airplane under complete control. Depending upon load, load distribution, and power the remaining engines are producing, critical 3-engine speed is 110-120 mph.

Maintain directional control by opposite rudder and aileron to bring the dead wing up. Bring airplane to a **shallow** climbing attitude and allow airspeed to increase. At same time, call for wheels up. Do not bother with trim now; that can be taken care of later. The airplane will climb on 3 engines if proper flying technique is followed.

Remember

1. Retract landing gear as soon as you are airborne.
2. Attain critical speed by lowering the nose.
3. Use rudder and minimum aileron to bring dead wing slightly above horizontal.
4. Determine which is the faulty engine. Decide whether to use it or feather it.
5. Complete 3-engine takeoff procedure. Maintain proper engine operating conditions by use of cowl flaps and correct power settings.
6. After recovery and climb, use only as much power as you actually need. Overboosting the good engines may put them within the detonation range and lead to early engine failure.

2-ENGINE FAILURE ON TAKEOFF

Failure of 2 engines on takeoff requires the pilot's and copilot's closest cooperation, but recovery can be successfully accomplished with the proper technique.

If the engine failure occurs during the takeoff run and enough runway remains, close throttles and bring the airplane to a stop.

Remember there are certain limitations below which the airplane will not fly.

1. There is the critical speed of 115 to 125 mph below which the airplane will not sustain flight. This speed is governed by load, distribution of load and how much power the good engines will deliver.

2. The airplane will not accelerate on 2 engines at or below the critical speed regardless of how much additional power is applied.

Recovery can only be effected after critical speed has been reached by nosing the plane down sharply, applying full power and raising the wheels if it has not already been done, picking up airspeed and raising dead wing to establish directional control as soon as possible. If plane will not hold a constant airspeed above critical airspeed, it indicates that the plane will not climb with the load on board.

It is difficult for a pilot to bring himself to nose down an airplane with only 200-300 feet of altitude available, but you must realize that this is the only possible way to save the airplane.

It is imperative to have all movable loads as near CG as possible.

Recovery

1. Apply rudder, aileron and forward stick until dead wing is well above the horizon, and the nose slightly below the horizon.

2. Apply full power on the good engines. Pilot opens throttles; copilot places propellers in full high rpm, landing gear switch "UP."

3. Do not feather the faulty engines unless you are absolutely sure they will deliver no power and are only creating more drag. **Be sure you know which engines are faulty and feather the right ones.**

4. Do not bother with trim until recovery is fully accomplished.

5. Do not try to climb before recovery is fully accomplished. Even though you have recovered successfully, you still stand a chance of losing the airplane unless you attain your **critical** airspeed before beginning the climb.

6. Use a minimum of aileron. Use no aileron at all until critical speed has been attained, unless absolutely imperative. Remember ailerons set up a flap action. The cleaner the wings, the more rapid the recovery. Generally rudder alone will effect the desired correction.

7. Keep a close watch on pressures and temperatures of the good engines, and adjust temperatures with cowl flaps.

Caution

Keep in mind that **stall plus yaw invariably equals spin.** With 2 engines out on one side the airplane will always tend to yaw. Therefore, avoid low speeds.

Maintain speed by holding a safe attitude.

Next in importance is your altitude. If necessary, sacrifice altitude for safe airspeed and attitude.

GO-AROUND WITH ONE ENGINE OUT

While making an emergency 3-engine landing, there may be an occasion when you will be forced to go around.

Follow this procedure:

1. The pilot calls: "Check high rpm." The copilot checks high rpm, and stands by to raise flaps.

2. The pilot opens throttles on the 3 good engines simultaneously, maintaining directional control with rudder, and keeping the dead engine wing slightly above the horizon.

3. Pilot calls: "Flaps up"; copilot places flap switch in "UP" position. Pilot calls: "Wheels up," when sure that contact will not be made with runway.

4. Do not reduce power until safe airspeed and altitude are attained.

Caution

Raise flaps immediately upon application of power. Do not wait until a safe airspeed is reached. You will not reach a safe airspeed with flaps down and only 3 engines operating.

Do not try to get directional control by differentiating throttles. **Open throttles on good engines fully.**

There will be a settling effect caused by loss of lift as the flaps go from ⅓ down to the full up position. Apply slight back pressure on wheel to increase angle of attack and thus compensate for this loss of lift.

2-ENGINE LANDING

A 2-engine landing will require a technique considerably different from that used for a normal landing.

Make a recovery as outlined under "2-Engine Failure on Takeoff" (p. 145). Then be sure that: (1) recovery is fully accomplished, (2) airspeed is safely above critical speed, (3) the airplane is well under control, and (4) you will not put yourself in a situation where a go-around is necessary.

WHEELS DOWN

SHALLOW TURN

AIRSPEED OVER 130 MPH

LOWER HALF FLAPS
SAVE THE OTHER HALF UNTIL LANDING IS IN THE BAG

CLOSE THROTTLES COMPLETELY BEFORE LANDING

Approach and Landing

1. Plan your approach so that a shallow turn toward the runway can be started as soon as possible.

2. Set manifold pressure and rpm as required to sustain safe flight.

3. Put down landing gear on base leg if the base leg is close to the field; otherwise wait until you are close enough.

4. Approach the runway at a constant rate of descent with airspeed above 130 mph.

5. Lower ½ flaps when it is apparent runway will be reached, and at the same time reduce airspeed between 120-125 mph.

6. When you are sure that you will not undershoot the runway, lower the remainder of flaps, further reduce airspeed to not less than 115 mph. Close throttles completely before landing.

7. Do not attempt a low dragging approach. Neither direction nor altitude can be maintained with full flaps and wheels down when operating on 2 engines on one side, even with full power.

8. Do not put flaps down until a landing is in the bag. Remember that dragging up to the field or going around is virtually impossible with wheels and flaps down and only 2 engines operating.

SINGLE-ENGINE OPERATION

Never take it for granted that the B-17 will fly on one engine at any altitude or at any power setting.

If the external condition of the airplane is clean and the operating engine is in good condition, flight may be made for a limited distance. However, the power required is more than one engine can continue to develop indefinitely. Therefore, the crew must be prepared to make a landing when the single operating engine fails.

If you attempt single-engine operation, don't use flaps until the time of landing. Jettison all possible equipment and close all hatches and windows. Feather the 3 dead propellers and close the cowl flaps on the 3 dead engines. Airspeed should be approximately 120 mph, not lower. Have crew members take stations in the cockpit and radio room in order to obtain a normal center of gravity of 28%-30% MAC.

The power required to maintain level flight at 5000 feet for a 40,000-lb. gross weight airplane at 120 mph IAS is slightly less than 1000 thrust Hp. At lower speeds the power required for this gross weight is still lower. One engine may just be able to develop this required power at 5000 feet or lower at a power setting exceeding military power. If all loose guns and equipment are thrown overboard and the fuel is low, the gross weight may be reduced to approximately 40,000 lb.

The practical value of the above procedure is to enable you to prolong your glide and maintain more control of the airplane. Thus, you may be able to reach a field for a landing that might be impossible otherwise.

Remember that with engines feathered there is less drag than with engines idling.

WITH 3 ENGINES FEATHERED YOU WILL FLOAT FARTHER IN YOUR FLARE-OUT FOR LANDING THAN IS USUALLY EXPECTED.

HOW TO BAIL OUT OF THE B-17

When an emergency develops and it becomes necessary to abandon the airplane in flight, there is no time for confusion or second guessing. Procedure of the entire crew in bailing out of the airplane must be almost automatic. Each crew member must know (1) his duties, (2) through what hatch he is supposed to exit, and (3) how to bail out, open his parachute, and land. (See PIF 8-4-1.)

As airplane commander, your first responsibility is to be sure that your crew is thoroughly trained, by regular ground drill, in the proper procedure for bailing out of the B-17.

Before taking off on any flight make absolutely sure that:

1. An assigned parachute, properly fitted to the individual, is aboard the airplane for each person making the flight.

2. The assigned parachute is convenient to the normal position in the airplane occupied by the person to whom it is assigned.

3. Each person aboard (particularly if he is a passenger or a new crew member who has not taken part in your regular ground drill) is familiar with bailout signals, bailout procedure, and use of the parachute.

DUTIES OF THE CREW

The Airplane Commander

1. Notify crew to stand by to abandon ship. The bell signal consists of three short rings on alarm bell. At first alarm all crew members put on parachutes.

2. Notify crew to abandon ship. Bell signal consists of one long ring on alarm bell.

3. Check abandoning of airplane by crew members in nose.

4. Clear bomb bay of tanks and bombs, using emergency release handle.

5. Turn on autopilot.

6. Reduce airspeed if possible. Hold ship level.

Copilot's Duties

1. Assist airplane commander as directed.

Navigator's Duties

1. Determine position, if time permits.

2. Direct radio operator to send distress message, giving all pertinent information.

3. Stand by emergency exit in nose of airplane.

Bombardier's Duties

1. Assist navigator.

2. Stand by emergency exit in nose of airplane.

Engineer's Duties

1. Assist pilot as directed.

2. Notify pilot when crew in nose has abandoned the airplane.

3. Stand by to leave via bomb bay immediately after crew in nose has abandoned airplane.

Radio Operator's Duties

1. Find exact position from navigator.

2. Send distress call.

3. Stand by to leave via bomb bay.

Ball Turret Gunner's Duties

1. Stand by to leave via main entrance door, or most practical rear exit as occasion demands.

Tail Gunner's Duties

1. Stand by to leave via tail gunner's emergency exit.

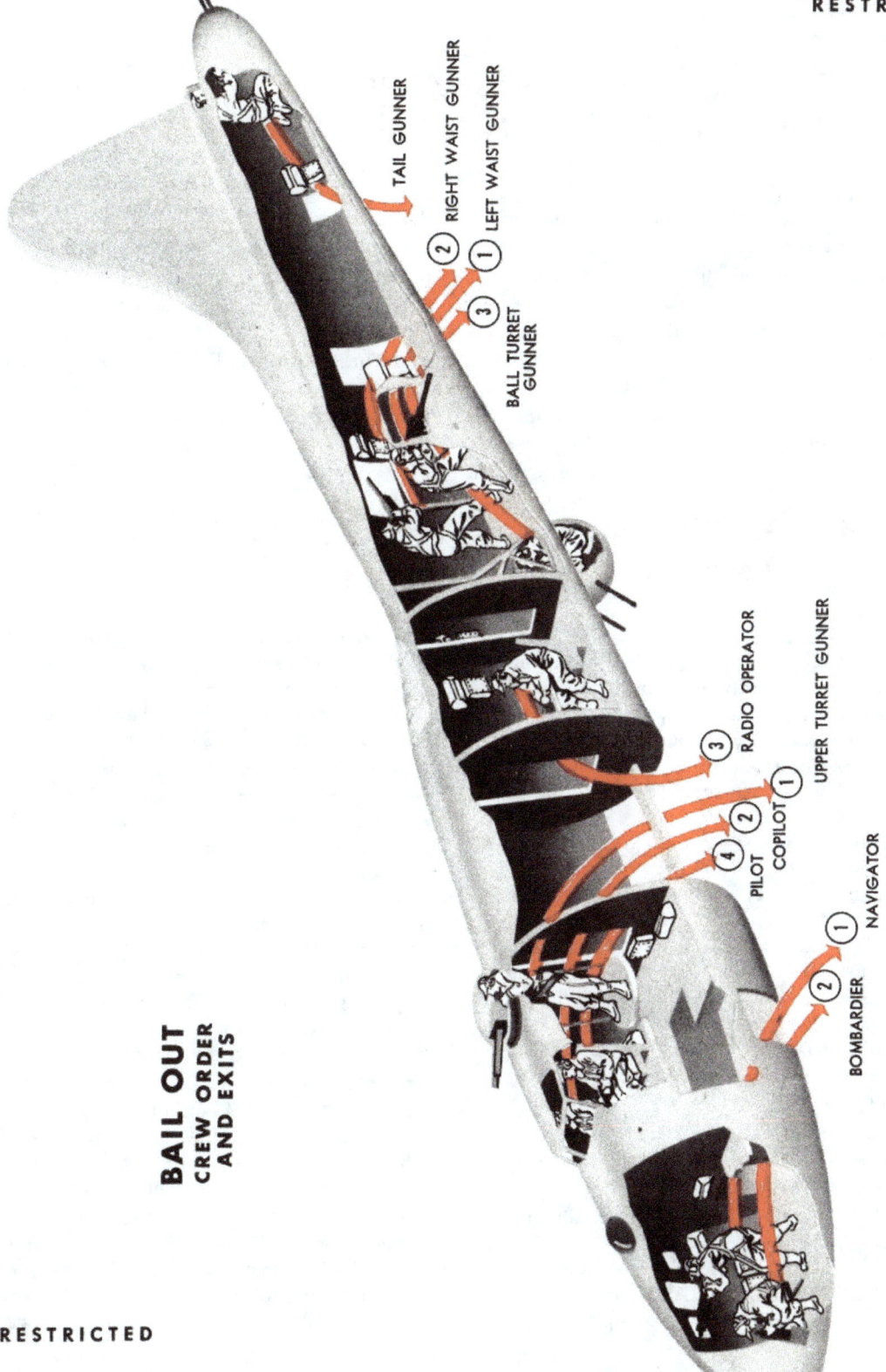

BAIL OUT
CREW ORDER AND EXITS

BAILOUT PROCEDURE WHEN WEARING THE CONVENTIONAL SEAT OR BACK-TYPE PARACHUTE

Pilot—Exits fourth out forward end of bomb bay. (Alternate exit, out front entrance door.) Is last to leave plane.

Copilot—Exits second through forward end of bomb bay.

Bombardier—Exits second through front entrance door.

Navigator—Exits first out of front entrance.

Upper Turret Gunner—Exits first out forward end of bomb bay.

Radio Operator—Exits third through after end of bomb bay.

Right Waist Gunner—Exits second through main entrance door.

Left Waist Gunner—Exits first out main entrance door.

Ball Turret Gunner—Exits third out of main entrance door.

Tail Gunner—Exits through small emergency door in tail.

BAILOUT PROCEDURE WHEN WEARING QUICK ATTACHABLE CHUTE HARNESS

When the order is given over the intercom to "Abandon airplane," each crew member will remove the individual seat-type dinghy and breast-type parachute from their respective positions near his station, snap them onto his QAC harness, and exit through the hatch specified. The following instructions, used with the diagram, show the positions of the dinghies and the parachutes, the correct exit hatch, and the order of bailing out. Where several crew members bail out of the same hatch, each should check the others to make sure that all are wearing a full complement of equipment, securely fastened. Whenever possible, jump from the after end of the hatch. Remember, a life vest should be worn under the QAC harness on all over-water flights. The lanyard on the dinghy should be snapped onto the D-ring on the life vest.

Periodic ground drills will familiarize your crew members with the operation of the QAC harness and the order of bailout.

Pilot—Parachute mounted on floor, directly behind pilot's seat in pilot's cabin. Dinghy worn in seat position. Pilot is fourth to exit through forward end of bomb bay. (Alternate exit, out front entrance door.) Last to leave plane.

Copilot—Parachute mounted on floor directly behind copilot's seat in pilot's compartment. Dinghy worn in seat position. Exits second through forward end of bomb bay. (Alternate exit, through front entrance door.)

Bombardier—Parachute mounted in navigator's compartment on starboard wall directly opposite navigator about halfway up on wall. Dinghy mounted in navigator's compartment near floor on starboard side, half the distance forward from bulkhead. Exits second through front entrance door.

Navigator—Parachute mounted on bulkhead armor plating directly above door, on inner side of navigator's compartment. Dinghy mounted alongside and to rear of bombardier's dinghy. Exits first through front entrance door.

Upper Turret Gunner—Parachute mounted on floor just forward of bomb bay bulkhead on port side. Dinghy mounted on forward wall of bomb bay bulkhead in turret compartment, directly below entrance to bomb bay. Exits first through forward end of bomb bay.

Radio Operator—Parachute mounted on starboard wall just forward of rear bulkhead of radio compartment, three-quarters of the way up side of wall. Dinghy mounted directly beneath parachute. Exits third through after end of bomb bay.

Right Waist Gunner—Parachute mounted on starboard wall just forward of rear door and even with top of door. Dinghy mounted directly beneath parachute. Exits second through main entrance door.

Left Waist Gunner—Parachute mounted on wall immediately aft and on same level as left waist window on port side. Dinghy mounted directly beneath parachute. Exits first through main entrance door.

Ball Turret Gunner—Parachute mounted on aft starboard side of rear bulkhead of radio compartment, about even with top of door. Dinghy mounted directly beneath parachute. Exits third through main entrance.

Tail Gunner—Parachute mounted on starboard wall immediately aft and slightly above rear gunner's escape hatch. Dinghy mounted directly beneath parachute. Exits through small emergency door in tail.

Wherever possible, jump from the after end of the hatch. Where several crew members bail out of the same exit, each should inspect the others to make sure that all are wearing a full complement of equipment, securely fastened.

Any other crew member, waist gunners and passengers will leave via main entrance door or most practical rear exits, as occasion demands.

Practice Bailout Procedure

After explanation of procedure, have the crew go to the airplane and practice abandoning airplane on the ground. Too much emphasis cannot be placed on the proper procedure, and on every man knowing his exit.

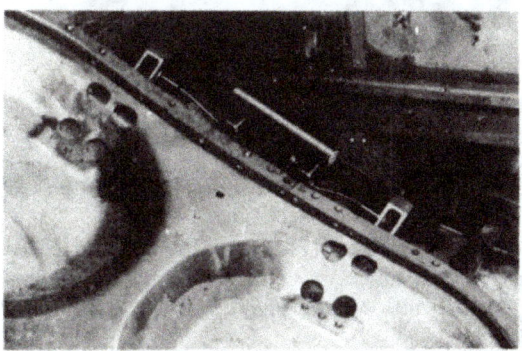

Emergency Release: Navigator's Hatch

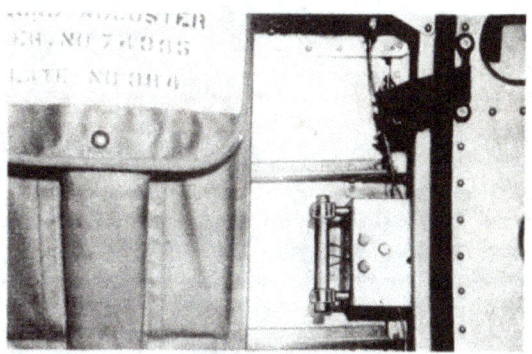

Emergency Release: Waist Door

Emergency Release: Tail Gunner's Hatch

BUILD CONFIDENCE AND CLEAR KNOWLEDGE BY HOLDING REGULAR BAILOUT DRILL

RESTRICTED

HOW TO DITCH THE B-17

Ditching drill is the responsibility of the pilot. Duties should be studied, altered if necessary to agree with any modifications, memorized, and practiced until each member of the crew performs them instinctively.

The pilot's warning to prepare for ditching should be acknowledged by the crew in the order given here—copilot, navigator, bombardier, flight engineer, radio operator, ball turret gunner, right waist gunner, left waist gunner, and tail gunner, i.e., "Copilot ditching," "Navigator ditching," etc.

Upon acknowledgment, crew members remove parachutes, loosen shirt collars and remove ties and oxygen masks unless above 12,000 feet. When preparations for ditching are begun above 12,000 feet, main oxygen supply or emergency oxygen bottle is used until notification by the pilot. All crew members wearing winter flying boots should remove them. No other clothing should be removed.

Releases on life rafts should not be pulled until the plane comes to rest.

Beware of puncturing rafts on wing and horizontal surfaces after launching. The dinghies should be tied together as soon as possible.

Injured men should get first consideration when leaving the airplane.

Life vests should not be inflated inside the plane unless the crew member is certain that the escape hatch through which he will exit is large enough to accommodate him with the vest inflated.

When personnel are in dinghy, stock of rations and equipment should be taken by the airplane commander (or copilot). Strict rationing must be maintained. Flares should be used sparingly and only if there is a reasonable chance that they will be seen by ships or aircraft. Don't forget the Very pistol.

Lash the life rafts together.

Landing crosswind is recommended unless the wind exceeds about 30 mph, in which case and into the wind. In executing the crosswind landing, the pilot will line up with the lines of the crests, at any convenient altitude, adjust flaps, power settings, trim, and make the approach with a minimum rate of descent, with a minimum forward speed. Land on a crest parallel to the line of crests or troughs. Crabbing will be necessary to remain over the crest while making the approach.

DUTIES OF THE CREW

Airplane Commander

(1) Give "Prepare for ditching" warning over interphone; give altitude; sound ditching bell signal of six short rings.

(2) Fasten safety harness.

(3) Open and close window to insure freedom of movement. Place ax handy for use in case of possible jamming.

(4) Order radio operator to ditching post.

(5) Order tail gunner to lower the tailwheel by cranking about 10 turns.

(6) 20 seconds before impact, order the crew to "brace for ditching." Give long ring on signal bell.

(7) Release safety harness and parachute straps. Exit through side window when airplane comes to rest. Inflate life vest.

(8) Proceed to left dinghy, cut tie ropes. Take command.

Copilot

(1) Assists pilot to fasten safety harness.

(2) Fastens own safety harness, opens and closes right window to insure freedom of movement.

(3) Releases safety harness, parachute straps, exits through right window when plane comes to rest. Inflates life vest.

(4) Proceeds to right dinghy, cuts ropes. Takes command.

Navigator

(1) Calculates position, course, speed, giving this information to the radio operator. Destroys secret papers. Gathers maps and celestial equipment. Gives wind and direction to the pilot.

(2) Proceeds to radio compartment. Closes radio compartment door.

(3) Attaches rope on emergency radio equipment and signal set (if radio is stored in radio compartment).

(4) Assumes ditching position.

(5) Hands the following items in the order given to the bombardier, who is already out: signal set and emergency radio, ration kits, navigation kits, parachutes.

(6) Exits through radio hatch and goes to left dinghy.

Bombardier

(1) Jettisons bombs, closes bomb bay doors, destroys bombsight, goes to radio compartment, closing compartment door. Takes first-aid kits to radio compartment.

(2) Takes position, partially inflates life vest by pulling cord on one side.

(3) Directs and assists exit of men through radio hatch. Stands above and forward of hatch and receives equipment from navigator and hands it to crew members as follows: signal set and radio to radio operator; ration kit No. 1 to tail gunner; ration kit No. 2 to right waist gunner; navigation kit to ball turret gunner; pigeon crate to left waist gunner. Assists flight engineer in making exit.

(4) Goes to right dinghy.

Flight Engineer

(1) Jettisons ammunition and loose equipment, turns top turret guns to depressed position pointing forward.

(2) Goes to radio compartment. Lowers the radio hatch and moves it to the rear of the plane, jettisons loose equipment in radio compartment, and slides back top gun.

(3) Stands with back to aft door of radio compartment and assists other members out by boosting them.

(4) Last man to leave radio compartment, with bombardier's help. Goes to left dinghy.

Radio Operator

(1) Switches on liaison transmitter (tuned to MFDF) sends SOS, position and call sign continuously, turns IFF to distress, remains on intercom, transmits all information given by navigator.

(2) Obtains MFDF fix, continues SOS, remains on intercom.

(3) On pilot's order clamps key, takes ditching position, inflating life vest partially, remains on intercom, repeating pilot's "Brace for ditching" to crew.

(4) Receives signal kit and emergency radio from bombardier.

(5) Assists with dinghy inflation and inspects for leaks.

(6) Goes to right dinghy.

Ball Turret Gunner

(1) Turns turret guns aft, closes turret tightly, goes to radio compartment with first-aid kits and ration kits.

(2) Pulls both dinghy releases as aircraft comes to rest.

(3) Goes to left dinghy.

Right Waist Gunner

(1) Jettisons his gun, ammunition, all loose equipment.

(2) Closes right waist window tightly, goes to radio compartment, collecting emergency radio and signal box in fuselage (if radio is stored elsewhere than in radio compartment).

(3) Takes position, partially inflates vest.

(4) Assists in inflating right dinghy, inspects for leaks, applying stoppers if necessary.

Left Waist Gunner

(1) Jettisons his gun, ammunition, loose equipment, closes left waist window, goes to radio compartment.

(2) Partially inflates vest.

(3) Receives pigeon crate from bombardier.

(4) Goes to right dinghy.

Tail Gunner

(1) Jettisons ammunition; goes forward, cranks down tailwheel about 10 turns; collects emergency ration pack (stowed in fuselage); is last to enter radio compartment.

(2) Takes position, partially inflates life vest.

(3) Carrying ration pack, goes to left dinghy, assists with dinghy inflation, inspects for leaks.

CREW POSITIONS FOR DITCHING

The positions illustrated should best enable crew members to withstand the impact of crash landings on either land or water. On water 2 impacts will be felt, the first a mild jolt when the tail strikes, the second a severe shock when the nose strikes the water. Positions should be maintained until the aircraft comes to rest. Study them carefully.

Emergency equipment for use in the dinghy should be carried to crash positions. Any equipment carried free must be held securely during ditching to prevent injury.

Parachute pads, seat cushions, etc., should be used to protect the face, head, and back.

1. Jettison bombs, ammunition, guns and all loose equipment and secure that equipment which might cause injury. Close bomb bay doors and lower hatches. If there is not enough time to release bombs or depth charges place them on "SAFE." Retain enough fuel to make a power landing.

2. Navigator calculates position, course, and speed and passes data to radio operator. Latter tunes liaison transmitter to MFDF and sends SOS, position and call sign continuously. Radio operator also turns IFF to distress and remains on intercom; clamps down key on order to take ditching post.

3. These tips will help you determine wind direction and speed: (a) waves in open sea move downwind; (b) direction of spray indicates wind direction; (c) wind lanes—a series of lines or alternate strips of light and shade— also show direction; (d) approach on waves should be made into wind at right angles to them; (e) approach on swells should be made along top, parallel to swell, and may be executed in winds not over 10 mph.

JETTISON LOAD....

SOS... AND ASSUME POSITIONS

BRACE FOR DITCHING

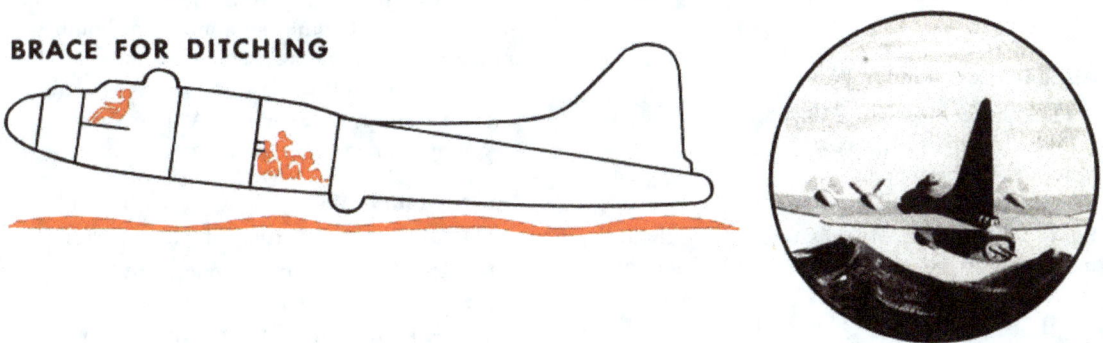

How to Determine Wind Speed

A FEW WHITE CRESTS	10 to 20 mph
MANY WHITE CRESTS	20 to 30 mph
FOAM STREAKS ON WATER	30 to 40 mph
SPRAY FROM CRESTS	40 to 50 mph

CRASH LANDINGS

No procedure can be established which will fit all cases. The following is a summary of the steps which should be taken if time permits. The airplane commander will exercise his authority to alter this procedure wherever necessary.

Airplane Commander Will

(1) Notify crew by interphone or oral communication between crew members that crash landing will be made.

(2) Notify bombardier to release bombs or bomb bay tanks. (If possible, drop them in uninhabited or enemy territory.) Then close the bomb bay doors.

(3) Make a normal slow landing, with flaps down and landing gear up.

The Copilot Will

(1) Turn master switch and battery switches "OFF" after operation of necessary electrical equipment such as flaps, radio, gear, landing lights, etc., when it is certain that there will be no further need for the operating engines.

(2) Assist airplane commander as directed.

The Bombardier Will

(1) Check with airplane commander to determine if auxiliary gas and/or bombs are to be dropped.

(2) Release bombs or tanks. Close bomb bay doors.

(3) Proceed to radio compartment.

The Engineer Will

(1) See that each enlisted man in the radio compartment is properly braced for impact.

(2) See that doors from radio compartment of airplane into bomb bay, and from bomb bay into control cabin are locked open.

(3) See that all emergency exits are opened, but not freed from airplane. A door that is cast free may damage the control surfaces.

The Navigator Will

(1) Determine position if time permits.

(2) Proceed to rear compartment.

(3) Direct radio operator to send distress message, giving all pertinent information.

Abandoning Airplane Following Crash Landing on Land

(1) All preparation for abandoning ship has been made during the approach. After landing, little can be done except to get out as quickly as possible.

(2) Crew members will take fire extinguishers, if available, with them when leaving the airplane. This may enable them to put out a small fire and rescue personnel trapped in the airplane.

(3) Dispose of all classified material in accordance with Army Regulation 380-5.

EQUIPMENT

Fuel System

The fuel system of the B-17F consists of 4 independent fuel supplies of approximately equal capacities, each feeding one engine. There are 3 tanks in each wing, with provisions for 2 additional groups of outer wing feeder tanks. These outer wing feeder tanks (Tokyo tanks) are composed of 9 individual, collapsible self-sealing cells per wing. The fuel supply can also be increased by auxiliary installations of releasable fuel tanks in the bomb bay.

The fuel in any tank is available to any engine supply tank in the airplane through a **fuel transfer system** consisting of 2 **selector valves** and an **electrical transfer pump**.

There is also a **hand transfer pump** in the bomb bay as an emergency transfer medium. **Fuel booster pumps** in the outlets of the 4 major wing tanks eliminate vapor lock between the tank and the engine fuel pump. They also provide fuel to the carburetor when the engine pump fails. An electrically controlled **fuel shut-off valve** is installed in the line beyond the fuel booster pump to prevent fuel flow through a severed fuel line.

FUEL CAPACITY

FUEL TANKS	U.S. GALLONS EACH	TOTAL U.S. GALLONS
No. 1 and No. 4 engines	425	850
No. 2 and No. 3 engines	213	426
Feeders (2)	212	424
Outboard Wing 1-5 (Total)	270	540
Inboard Wing 6-9 (Total)	270	540
Total Fuel (Overload)		2780
Bomb Bay Extras (2)	410	820
Total Fuel (Special)		3600

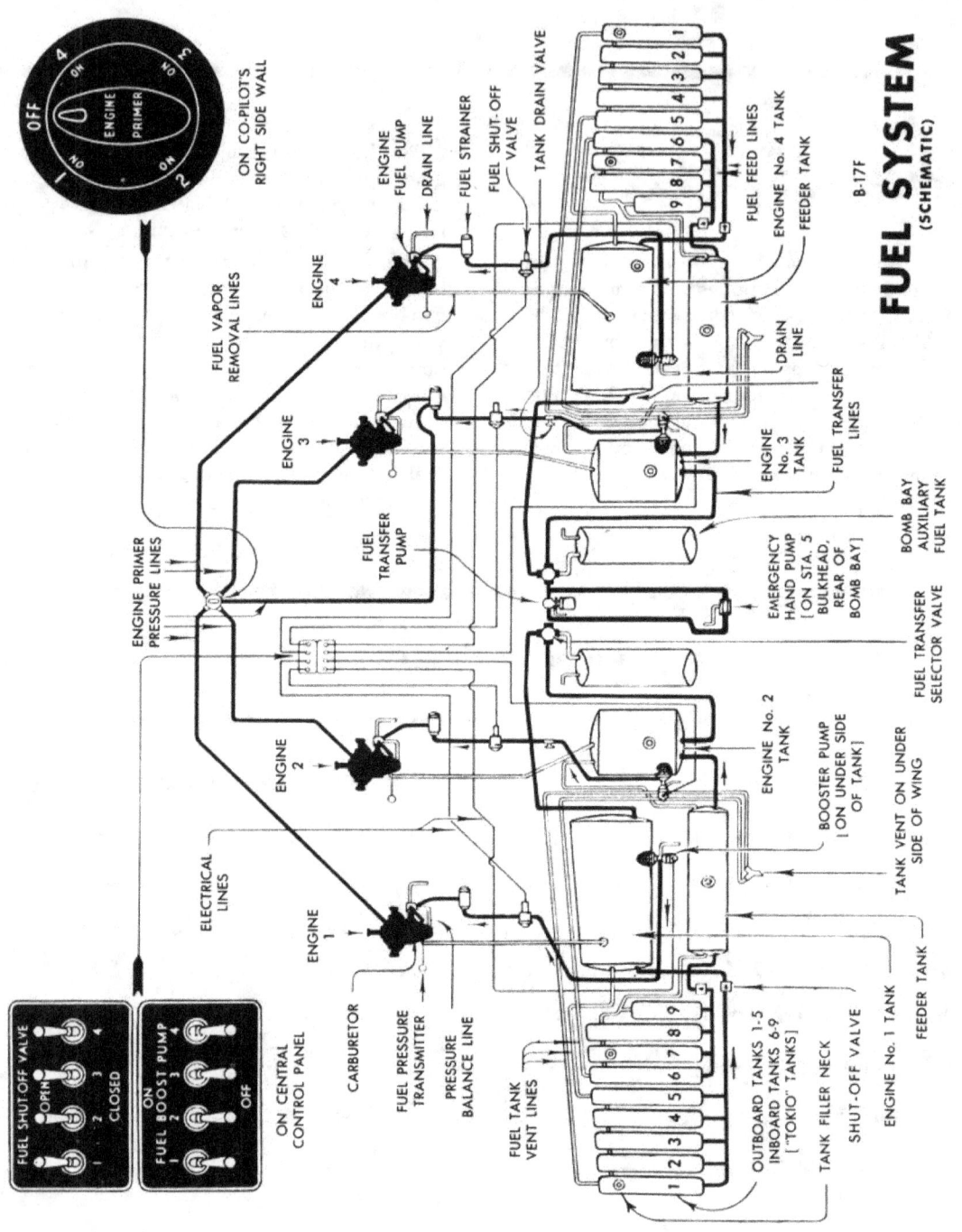

Booster Pump

The booster pumps (at the outlet on the underside of each of the 4 main tanks) serve to: (1) assure fuel to the engine fuel pump on take-off and landing, and when flying at less than 1000 feet or more than 10,000 feet above the ground; (2) prevent vapor lock in the fuel lines; (3) provide fuel to the carburetors when starting engines. No. 3 booster pump also supplies pressure to the primer pump at engine starting.

They are electrically operated and controlled by toggle switches on the central control stand. At high altitudes, bubbles form in the gasoline. As the gasoline is drawn through the funnel, the centrifugal action of the propeller throws these bubbles out through the sides of the screen, back into the tank. Only the liquid gasoline enters the pump and is sent to the fuel system.

Turn booster pumps on below 1000 feet; turn them on above 10,000 feet as a safeguard against vaporization.

Fuel Shut-off Valves

Shut-off valves provide an emergency means of shutting off fuel flow in case the fuel lines are severed. Valves for tanks No. 1 and No. 4 are forward of the tanks between the oil coolers. Valves for tanks No. 2 and No. 3 are between the tanks and the rear spar. Each valve is spring-loaded to stay open and is closed by means of a solenoid controlled by an individual toggle switch in the cockpit.

Engine-Driven Fuel Pump

The fuel pump forces sufficient fuel to the engines for operation at altitudes up to 10,000 feet. Above 10,000 feet, the fuel pump must be assisted by the fuel booster pump located on the right-hand engine accessory pad.

Fuel is drawn into the pump by the paddle-wheel action of the vanes within the liner. Fuel caught between the vanes at the inlet port is forced between the inner wall of the liner and the rotor and is carried to the outlet port. When the pumped fuel is in excess of the carburetor's demand, the excess fuel has sufficient pressure to lift the pressure-regulating valve from its seat. This permits the excess fuel to escape to the inlet side of the pump.

Fuel used in starting is pumped by the booster pumps through the engine-driven fuel pump. The fuel enters the inlet port (the engine driven fuel pump is now idle) and forces the bypass valve open, which permits the starting fuel to flow through the engine-driven pump to the carburetor.

Engine Primer

Provides a means of priming the engines for starting. It is on the floor to the right of the co-pilot. Fuel is drawn into the primer from the nacelle No. 3 fuel strainer and is forced into the top 5 cylinders of the engine selected. Several strokes are usually necessary to draw the initial flow of fuel into the primer.

(See starting procedure, p. 57, for operating information.)

Fuel booster pump for No. 3 engine must be turned "ON" to operate the primer. Do not leave plunger of engine primer in the up position as this allows fuel to pass directly through the primer to the engine selected.

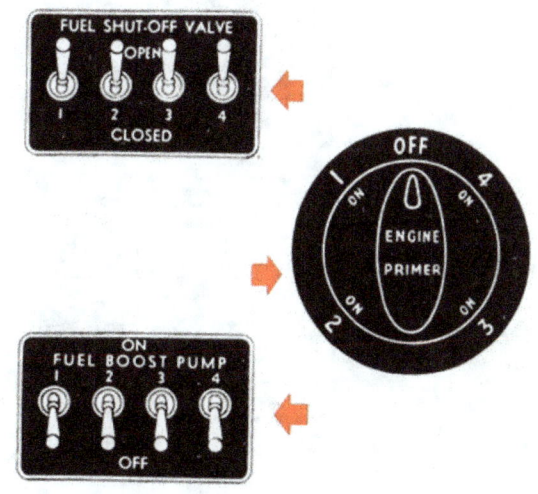

RESTRICTED

Fuel Transfer Selector Valves

Two selector valves direct fuel from any tank on one side of the airplane to any tank on the opposite side, exclusive of the inboard feeder tanks and the Tokyo tanks. To direct fuel from one tank to another **on the same side** of the airplane center line, the valves must first be set to transfer the fuel to a tank on the opposite side and then transfer it back across the center line to the desired tank.

The 2 selector valves are on the aft side of bulkhead in the rear of the pilot's compartment, one on the right and one on the left side of the door. The control handles are on the forward side of the bulkhead. When the shaft is turned, the cam also turns, and presses down one of the 3 plungers which open the desired valve. The fuel always enters the selector valve at one port and will exit from only one of the other 3 ports at one time.

An electric switch, installed as a safety feature on the handle of each of the 2 selector valves, closes the circuit to the pump motor whenever any valve port is opened. This eliminates any possible damage to the motor or selector valve in case all of the ports in one valve are closed.

1. Check the fuel transfer for proper operation at each preflight inspection.

2. Transfer fuel from the bomb bay tanks to the wing tanks as soon as possible after takeoff to check transfer system for operation. If you know that the transfer system is in operating condition, there is no need to hurry the transfer of fuel from bomb bay to wing tanks. Fuel in the bomb bay tank is **disposable load**—the most desirable kind of load to have if and when an emergency arises.

Fuel Transfer Pump

The fuel pump is used in conjunction with the transfer valves to transfer fuel from the

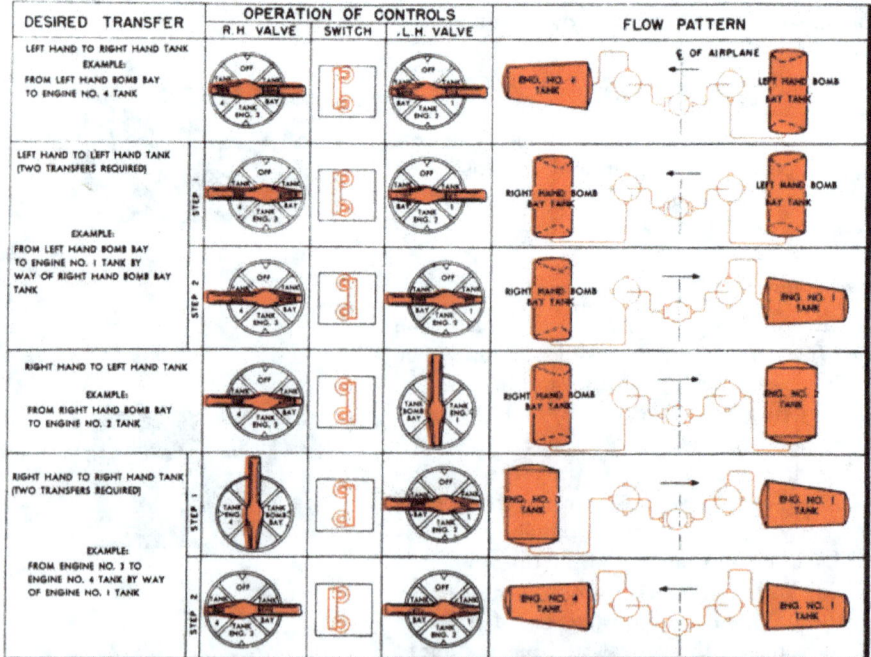

auxiliary tanks to the main wing tanks, or from one wing tank to another. It is in the forward end of the bomb bay under the step on the catwalk between the fuel transfer selector valves.

Hand Transfer Pump

A hand transfer pump on the rear bulkhead of the bomb bay provides a means of transferring fuel in case the electric-driven fuel transfer pump fails, or transferring fuel from drums to airplane tanks. The pump handle is turned in a clockwise direction. For transferring fuel in flight, disconnect the hose from the fuel transfer pump connection at fuel transfer selector valve. Connect the suction line of the hand pump to the selector valve on the side of the airplane from which the fuel is to be removed. Connect the pressure line of the hand pump to the selector valve on the side of the airplane to which the fuel is to be transferred.

Oil System

The oil system of the B-17F airplane has several functions: (1) it provides lubrication for wearing surfaces of the engine; (2) it aids as a coolant in transferring heat away from the engine; (3) it supplies hydraulic pressure to operate the supercharger regulation; (4) it supplies hydraulic pressure to operate the propeller pitch and propeller feathering mechanism.

Each engine has its own independent oil system. The self-sealing oil tanks are in the nacelles. The oil cooler and oil temperature regulators are in the leading edge of the wings. The hydraulic supercharger regulators are in the nacelles for the outboard engines and just aft of the superchargers for the inboard engines. The propeller feathering motors and pumps are on the forward side of each nacelle firewall.

Operation

Oil flows from the tank by gravity and by suction from the engine-driven oil pump, which forces the oil under pressure to the various moving parts of the engine. The oil then drops down to the sump, where it is picked up by the engine-driven scavenging pump and forced through the oil cooler. The oil then returns to the tank.

The oil lines to the supercharger regulators are tapped off the engine accessory cases on the pressure side of the pump. This oil circulates under pressure to the regulator and then returns to the engine, where it drains into the sump.

The propeller feathering oil line is tapped off the main oil line from the tank to the engine. The propeller feathering pump draws the oil from this line and forces it under pressure to the propeller feathering valve in the propeller dome.

All the oil lines are lagged (insulated) in order to prevent oil cooling and congealing at high altitude.

An oil dilution fuel line is tapped into the main oil line from the tank at the Y cock drain valve.

Oil Cooler

To cool engine oil returning from the crank case to the supply tank, there is an oil cooler for each engine. It consists of the core and muff and the oil temperature regulator.

The core passes the oil through a large cooling area; the muff is a bypass of the core in case the core becomes congealed.

The oil temperature regulator controls the amount of cooling air that passes through the core and is operated by the temperature and pressure of the engine oil.

Operation of the oil cooler shutters is fully automatic; therefore there are no oil cooler controls in the cockpit.

OIL SYSTEM DIAGRAM ON NEXT PAGE

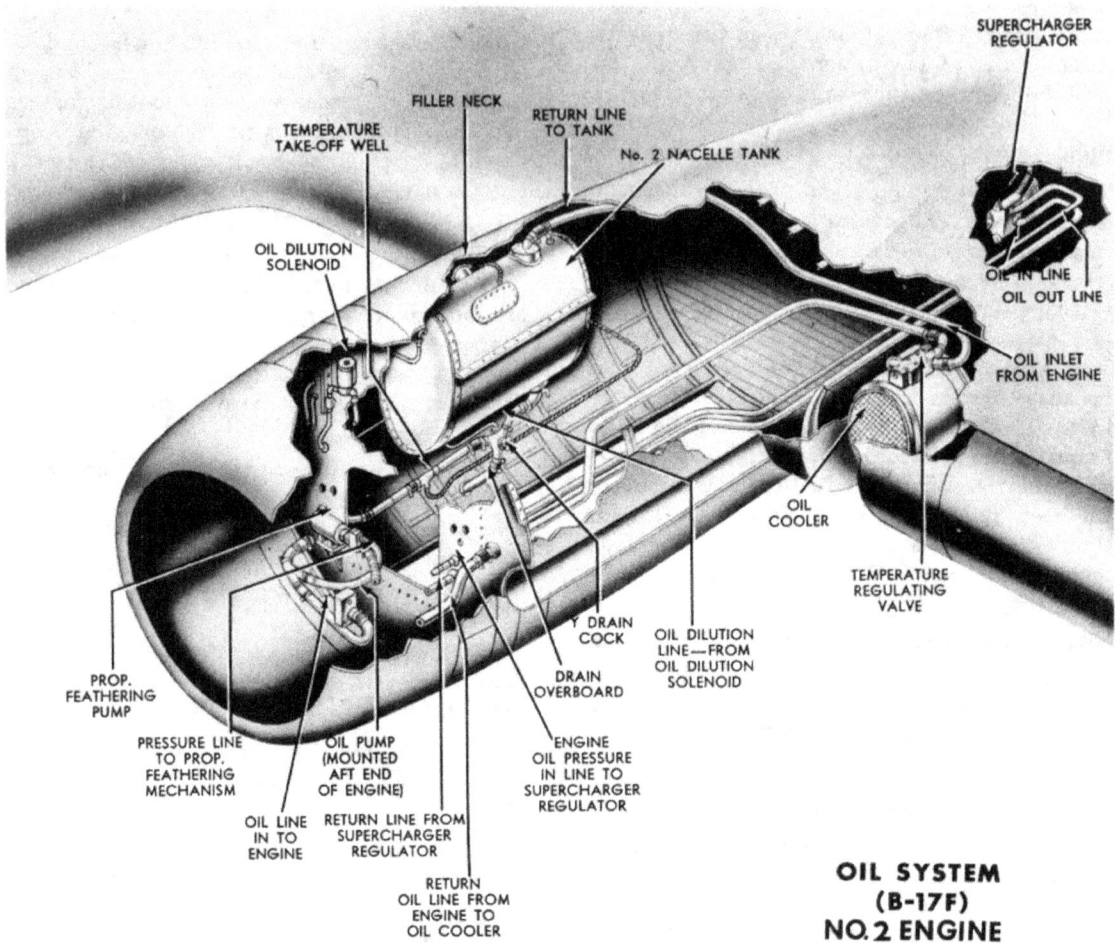

OIL SYSTEM (B-17F) NO. 2 ENGINE

Each engine is equipped with a self-sealing oil tank having a capacity of 37 gallons plus approximately 10 per cent expansion space.

The total of 148 gallons for all four tanks is required for maximum fuel load with wing tanks and bomb bay tanks full.

Fill oil tanks with Specification No. AN VV-O-446, grade 1120 for normal operations, grade 1100A for cold weather.

Hydraulic System (B-17F)

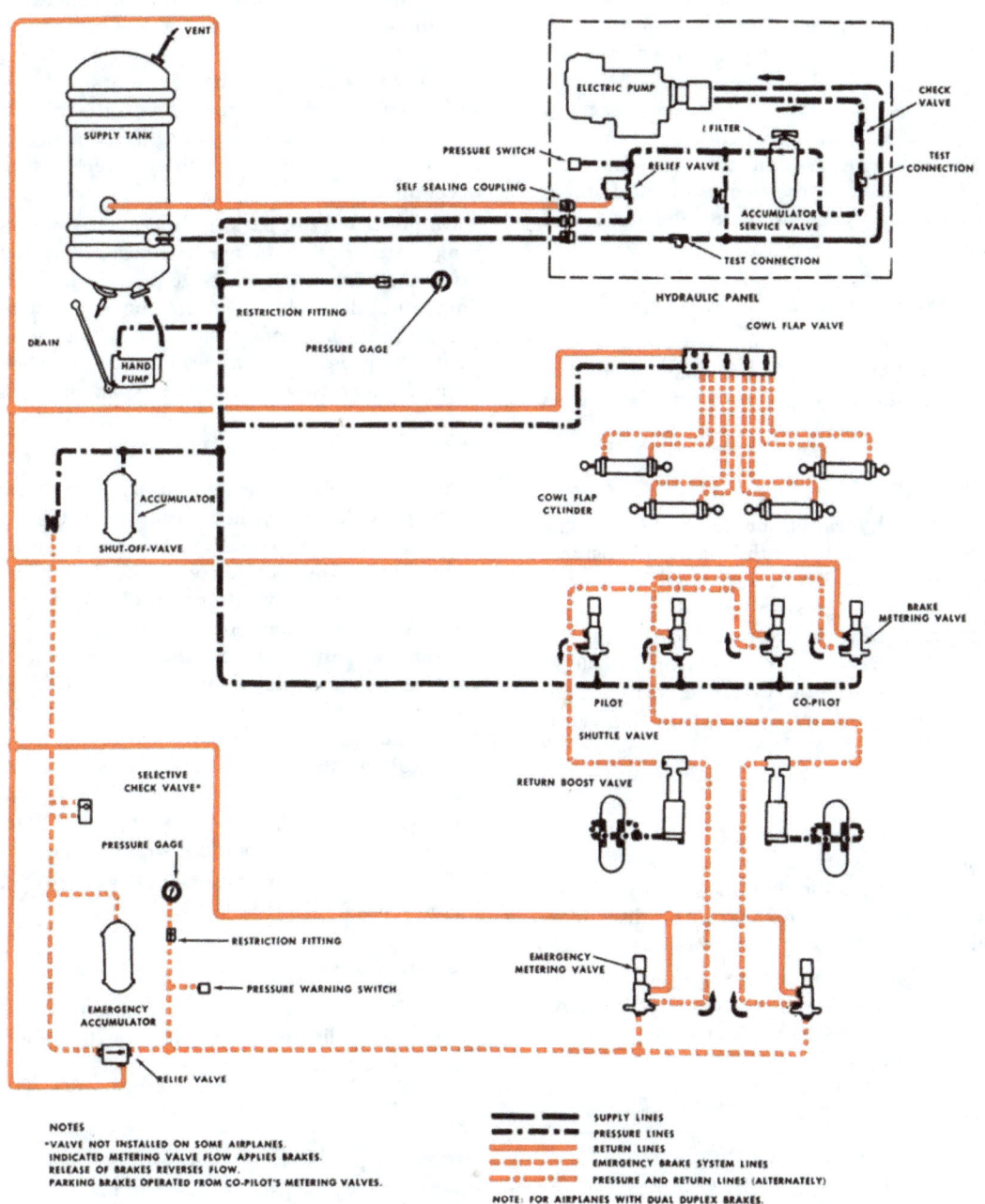

The hydraulic system on the B-17F operates the cowl flaps and the wheel brakes. It consists of a main system and an emergency system for operation of the cowl flaps and the brakes.

Operating pressures of the system are from 600 to 800 lb. sq. in. These pressures are developed by an electrically driven hydraulic pump which serves both the main and emergency systems. However, in all flight operations, the emergency system is shut off from the main system and relies on the hydraulic fluid stored in the emergency accumulator for its source of power.

System Oil and Capacity

The hydraulic oil used in the hydraulic system of the B-17F is AN VV-O-366a, and the total hydraulic oil capacity in the system is approximately 6 gallons.

Operation

When the hydraulic pump switch on the pilot's control panel is in the "AUTO" position, pressure is automatically regulated by a pressure cut-out switch, starting the pump when the pressure drops to 600 lb., and stopping the pump when the pressure builds up to 800 lb. In case the automatic pressure switch fails, maintain pressure by holding the hydraulic pump switch in the "MANUAL" position. A relieve valve opens if the pressure in the system reaches 900 lb.

Should leakage occur in the hydraulic system, stop the pump to prevent loss of fluid. Remove the hydraulic pump switch fuse in the main fuse panel in the cockpit, or disconnect the electrical receptacle at the pressure switch.

In some airplanes the hydraulic pump is controlled by an on-off switch on the pilot's control panel. This switch must be "ON" to maintain pressure.

Brakes

The brake assemblies are on the inboard side of the main landing wheels, except in B-17G and late F's, which have dual brakes.

Hydraulic pressure applied from the cockpit expands the expander tubes, forcing the brake lining against the brake.

Apply the brakes as little as possible and then only for short, hard intervals. Excessive and unnecessary use of the brakes will generate sufficient heat to cause failure of the expander tubes and cracking of the brake drums and wheels. Taxi the airplane with the inboard engines shut off and maintain directional control with the outboard engines when mission is completed.

Do not leave the parking brake on while the brakes are hot from previous use. This will cause the heat in the drums to pass through the lining and literally cook the expander tube, which then becomes brittle. Do not apply hydraulic pressure to the brake with the wheel removed, as this will burst the expander tubes.

Emergency Brake System

A spare accumulator and auxiliary metering valve provide emergency brake operation. A red warning lamp on the pilot's instrument panel lights when pressure in the emergency system falls to approximately 700 lb. sq. in. To charge the emergency accumulator, open the manual shut-off valve. If a selective check valve is installed, place it in the "SERVICING" position unless it is lock-wired in the "NORMAL" position. (These units are on the right side wall at the rear of the pilot's compartment.) Build up 800 lb. pressure in the system, then return the selective check valve to "NORMAL" and close the manual shut-off valve.

The emergency brake system has been eliminated from later-model airplanes.

Pressure Gages

Pressure in the service and emergency brake systems is indicated by 2 gages on the pilot's instrument panel.

Hand Pump

A hand pump on the side wall at the right of the copilot is used to supply pressure for ground operations and to recharge the accumulators if the electric pump fails.

Electrical System

Electrical power operates much of the auxiliary equipment in the airplane, such as the turrets, landing gear, wing flaps, instruments, bomb bay doors, and other miscellaneous equipment. Various units of the electrical system are distributed throughout the entire airplane. (See diagram.)

A 24-volt direct-current system is used in the B-17F. Type Mg-149 inverters are installed to furnish alternating current for all equipment requiring alternating current for its operation.

Control of the electrical system is accomplished mainly at the pilot's and copilot's stations. The bombardier and the navigator control the units necessary to their jobs.

Fuse shields, accessible in flight, are on the bulkhead to the rear of pilot's seat and the bulkhead in the radio compartment. There are also fuse shields in each nacelle. An alternating current fuse shield, accessible in flight, is on the floor below the pilot.

Generators

The generators on the accessory panel on the rear of each engine are the primary source of electrical power. They keep the batteries charged and provide power for electrical equipment while in flight. The generators are driven by the engines at 1½ times engine speed. They will deliver power at engine speeds above 1350 or 1400 rpm.

Auxiliary Power Equipment

A gasoline engine-driven generator unit, in the rear of the fuselage and for use only on the ground and in the air for emergencies, supplies auxiliary electric power for battery recharging or limited radio operation.

AC System

Alternating current for the autosyn instruments, drift meter, radio compass, and warning signals transformer is furnished by either of 2 inverters, one of which is a standby for the other. One inverter is under the pilot's seat and the other under the copilot's seat. A single-pole, double-throw switch on the pilot's control panel controls the DC power to the inverters and selects the inverter to be used. In the "NORMAL" position the left-hand inverter is on and in the "ALTERNATE" position the right-hand inverter is on.

Use of Auxiliary Power

Don't use engine generators in ground operation. Since it is inadvisable to deplete the batteries unnecessarily, another source of energy should be used in starting the engines.

Use the auxiliary power unit wherever practicable for ground operation. This not only saves the batteries but charges them, and use of this unit assures that it is in serviceable condition if it should be needed in emergency.

If you cannot use the auxiliary unit, start engines with battery carts or with a field energizer. Saving the batteries is especially important in preflight and cold weather.

Function of the Voltage Regulator

The engine generator, mounted in back of each engine, is geared to turn three times while the engine turns twice. The variable rpm of the engine would tend to vary the generator output were it not for the voltage regulator in the accessory section of the airplane.

The regulator operates by a variable resistance which changes the strength of the field magnets of the generator. The variable resistance is affected by an electromagnet which operates against spring tension. Voltage setting of the generator is set by varying the spring tension of the regulator or by varying the amount of current allowed to flow into the electromagnet, depending on the particular type of regulator used. Voltage regulators are in a shield under the pilot's floor in catwalk leading to bombardier's compartment.

Reverse Current Relay

A reverse current relay in each nacelle connects the positive lead of the generator to the power circuit bus in the back of the nacelle. This relay is usually set at 26½ volts. It cannot close unless the generator switch is closed, and it should automatically open whenever the current flow reverses (battery to generator). The possibility that this relay might stick and motorize the generator is another reason for not turning on the generators for ground operation.

Checking and Adjusting Generator Systems

Whenever starting engines, check the generators individually. After warming engines at 1000 to 1200 rpm, run up each engine slowly for check. Before running up an engine, turn on that particular generator. The pointer on the voltmeter will be somewhere in the center of the dial. The ammeter should not register, as voltage is too low to close the relay. As engine rpm is increased voltage increases. Between 26 and 27 volts, the ammeter should suddenly indicate amperage, showing that the relay has closed. Voltmeter should reach its maximum reading well below 1800 rpm. Turn off generator and repeat check with each of the other generators. **Only during the check should the generators be turned on individually.**

Equalizer Coils

Under ordinary conditions generators should be set to give a 28-volt reading on an accurate voltmeter. Sometimes voltage may vary one-half volt higher or lower. As long as all 4 generators maintain exactly the same voltage, amperage loads will be equal and the system is considered equalized or paralleled. A special equalizer coil is incorporated in the electromagnet of the voltage regulator and is interconnected with the equalizer coils of the other regulators. These coils help to maintain equal voltage and amperage of the 4 generators.

If in flight the ammeters show too great a disagreement, a paralleling adjustment is necessary. If one ammeter reads higher than the others, it is only because the voltage is a trifle higher on that generator. A slight adjustment of the voltage regulator by the flight engineer will correct this condition. The total output of the 4 generators remains the same; therefore, if the amperage of one is increased, the amperage of the other will be decreased.

When equalizing the generators it is advisable to synchronize the propellers and fly straight and level at a moderate cruising rpm—preferably about 1850 rpm. Leave all generators on, of course; otherwise current cannot flow from generators. Turn off batteries and all possible electrical equipment. The inverter alone will use enough current to cause all the ammeters to give a reading.

The less adjustment you make, the better. Careless adjustment may alter the voltage from the desired 28 volts. If ammeters read within 3 amperes with only inverter on, they will be within 20 amperes of each other at normal load.

Remember:

(1) Voltmeters and ammeters of the B-17 indicate only generator performance. If generators are not turning, these instruments do not function.

(2) The voltmeter reads any time the generator turns, whether generator switch is off or on.

(3) The ammeter indicates amount of current the generator is supplying. The more equipment in use, the higher it reads.

(4) Reverse current relays are not perfected. Therefore, don't use generators in ground operation when engine speed may not be great enough to keep all relays closed.

(5) As long as generators function properly, batteries will be charging; the batteries supply no current to electrical equipment while generator is on. Do not hesitate to turn off a battery if you believe it advisable. Sometimes you can save a boiling battery if you turn it off in time.

(6) Never turn off a good generator in the air, except perhaps momentarily to check another one. When a generator is left running and not putting out current, its commutator is apt to glaze and be inefficient. Also, the relay has a tendency to chatter, and its points may burn when this happens.

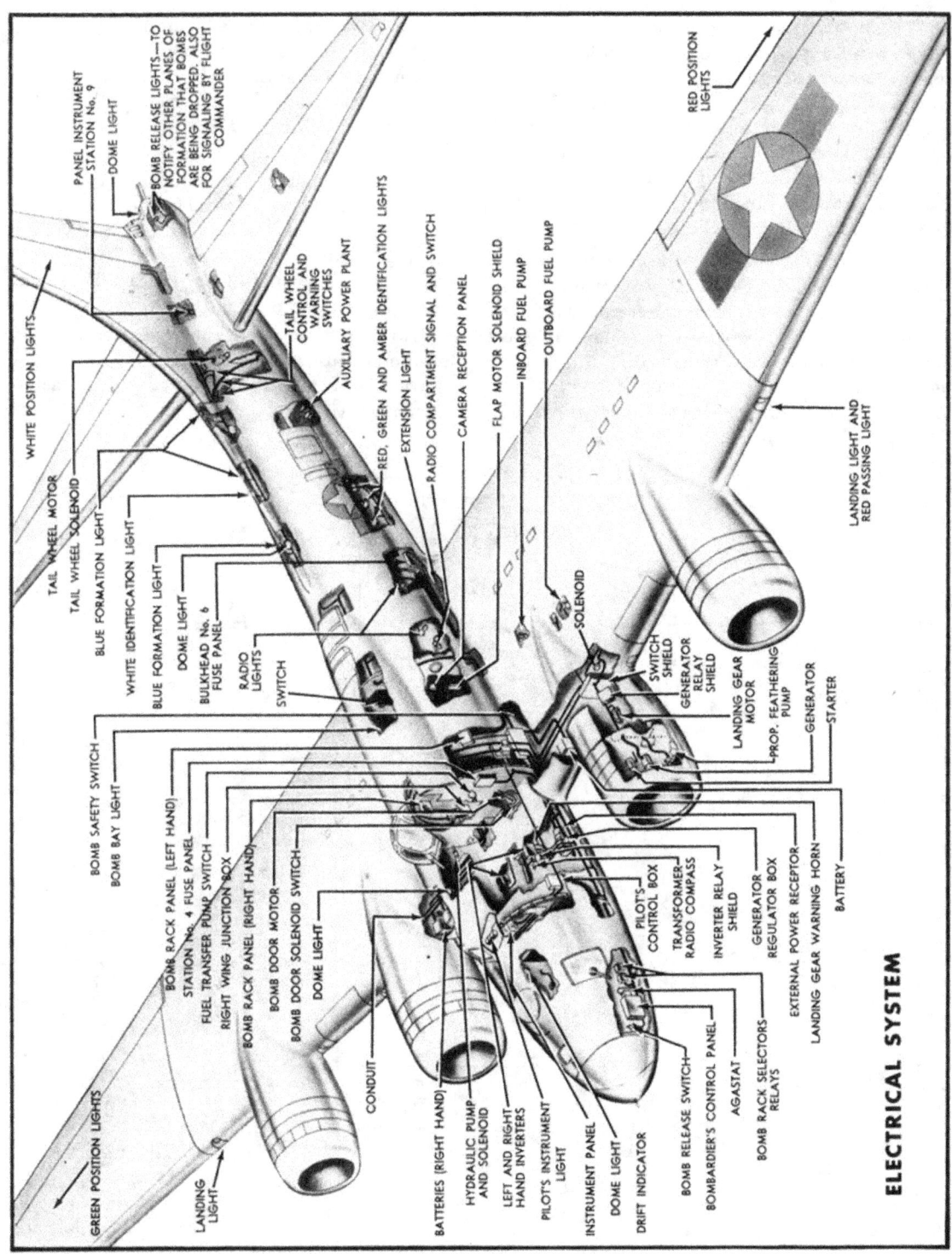

ELECTRICAL SYSTEM

RESTRICTED

(7) The purpose of the equalizing coil is to help equalize generator loads **only** when slight voltage variation causes unequal ammeter readings. If one generator is left off and the others are on and producing much current, too much load may be placed on the equalizer coil and the regulator may be damaged. Either have all **properly functioning** generators on or all off (except when checking).

(8) A bad generator is never completely disconnected from the electrical system until the regulator is removed. If a generator will not operate properly, remove its regulator. If you are flying, keep that switch off. If on the ground, remove the generator cannon plug also, to prevent wiring damage, and tape off the cockpit switch. In the B-17 the loss of a generator is not serious. More than twice as much power is available than will be needed. If a bad generator is properly disconnected the rest of the system will not suffer.

(9) Do not ask engineer to parallel generators when engines are operating at less than 1800 rpm. Don't try to parallel them on takeoff; wait until you have leveled off. Set voltage with all generator switches off. Do all your minor amperage paralleling with all switches on and little load on the system.

(10) The pilot **must** know location and disposition of fuses in fuse panels. Replacement of burned-out fuses often makes emergency action unnecessary.

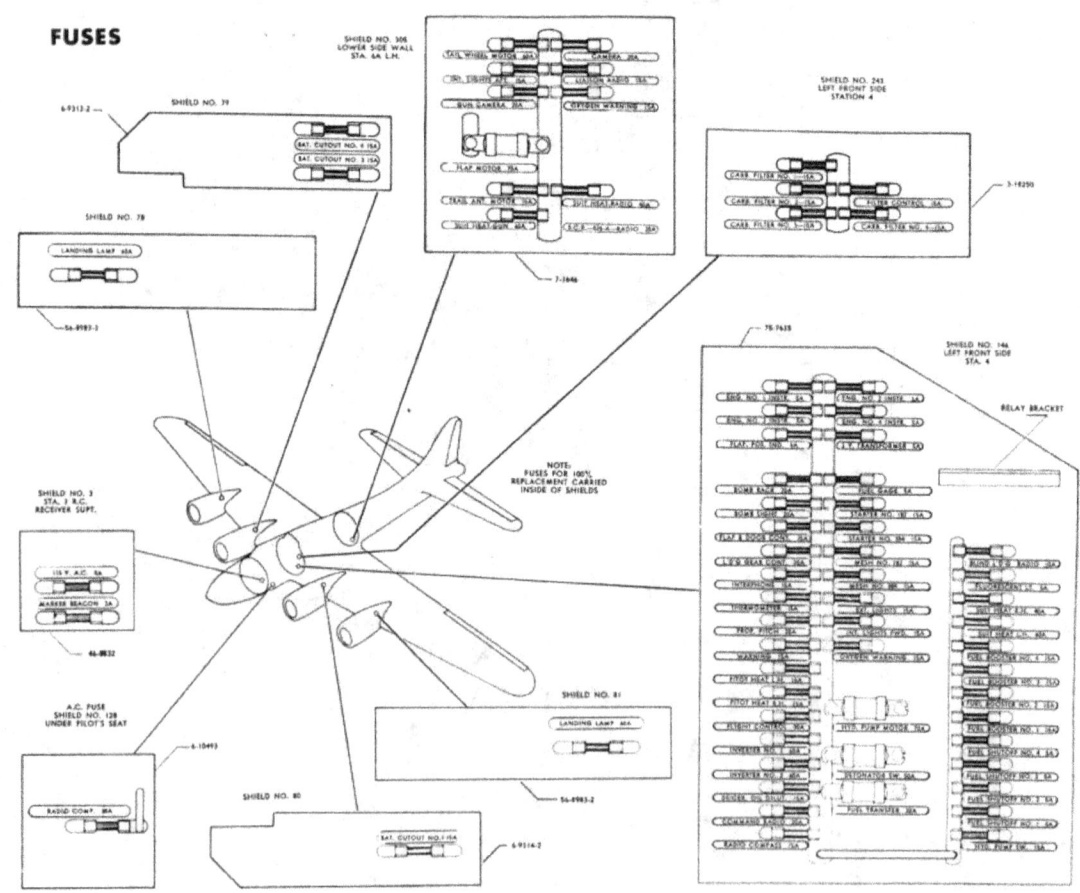

FUSES

Turbo-superchargers

Each engine on the B-17 has a turbo-supercharger which boosts manifold pressure for takeoff and provides sea-level air pressure at high altitudes.

To operate the turbo-superchargers, engine exhaust gas passes through the collector ring and tailstack to the nozzle box, expands to atmosphere through the turbine nozzle, and drives the bucket wheel at high speed.

A ramming air inlet duct from the leading edge of the wing supplies air to the impeller, which increases pressure and temperature. However, in order to avoid detonation at the carburetor, the air supplied to the carburetor passes through the intercooler, where the temperature is reduced. The internal engine impeller, driven by the engine crankshaft, again increases air pressure at it enters the intake manifold. Higher intake manifold pressure results in greater power output.

Supercharger Regulator Operation

The amount of turbo boost is determined by the speed of the turbo bucket wheel. Speed of the bucket wheel is determined by the pressure-temperature difference between the atmosphere and the exhaust in the tailstack, and by the amount of gas passing through the turbine nozzles. If the waste gate is open, more exhaust gas passes to the atmosphere via the waste pipe, decreasing the tailstack pressure.

The boost lever at the pilot's control stand sets the turbo regulator which automatically operates the waste gate to hold constant pressure in the tailstack. High boost lever setting provides higher exhaust manifold pressure by closing the waste gate. The resulting higher bucket-impeller speed gives higher intake manifold pressure.

The regulator (operated by engine oil) automatically opens and closes the waste gate to maintain a constant exhaust stack pressure equal to the boost lever setting.

Electronic Turbo-supercharger Control

The electronic turbo-supercharger control system on late model B-17's consists of 4 separate regulator systems, one for each engine, all simultaneously adjusted by a **manifold pressure** (turbo boost) **selector** dial on the pilot's control panel. Induction pressures are controlled through a Pressuretrol unit connected directly to the carburetor air intake.

Electrical power for the entire system is derived from the 115-volt, 400-cycle inverter.

Each regulator includes a turbo governor which prevents turbo overspeeding both at high altitude and during rapid throttle changes.

The exhaust waste gate is operated by a small reversible electric motor which automatically receives power from the regulator system when a change in waste gate setting becomes necessary to maintain desired manifold pressure.

In case of complete failure of the airplane electrical system, or failure of the inverter, the waste gates on all engines will remain in the same position as when failure occurred, and the same manifold pressure will be available as was in use at the time of failure.

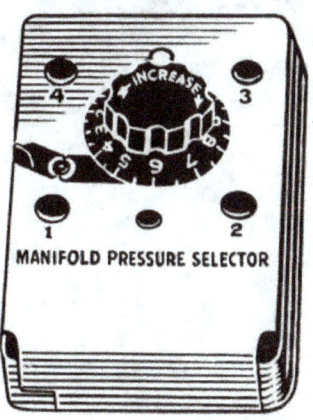

MANIFOLD PRESSURE SELECTOR

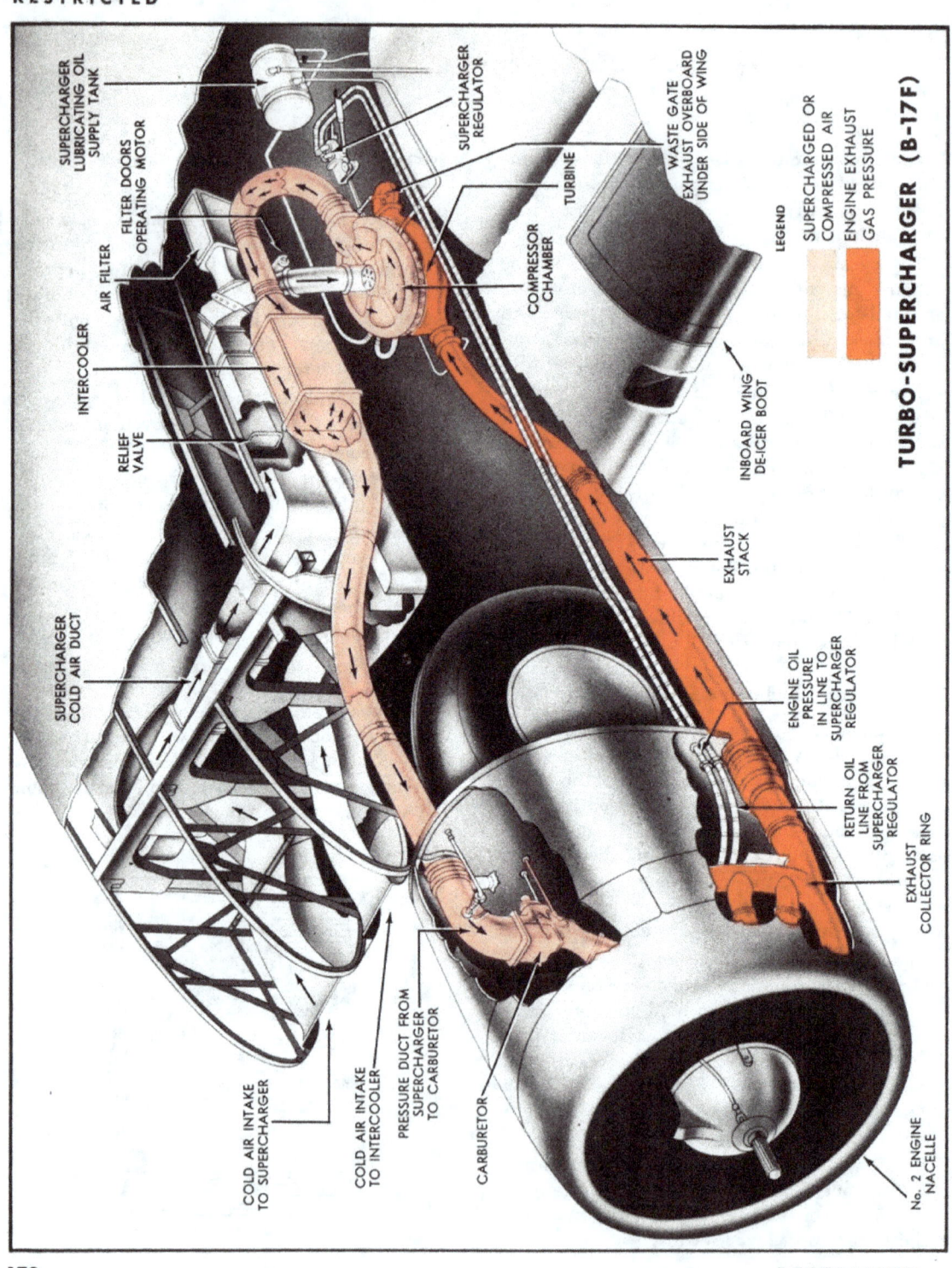

Operation: Before Starting

Set the turbo boost selector dial at zero to insure that the waste gate is open.

Taxi with the turbo boost selector dial at zero.

Engine Run-up

1. Set throttles at 1500 rpm on all engines.
2. Exercise propellers.
3. Check generators.
4. Electronic turbo control needs no exercising; therefore turbo boost selector remains on zero.
5. Return throttles to 1000 rpm on all 4 engines.

Before Takeoff

1. Run up engine No. 1 to 28" manifold pressure, and check the magneto.
2. Open No. 1 engine to full throttle, with the turbo boost selector on zero, to insure that waste gate is open.
3. Reduce No. 1 engine to 1000 rpm.

Follow the above procedure on engines No. 2, No. 3, and No. 4.

4. Having checked the magnetos, open engine No. 1 to full throttle, and turn the turbo boost selector dial clockwise until the desired takeoff manifold pressure is reached. (If the electronic control has been properly adjusted, you will obtain this manifold pressure at reference point "7" on the turbo boost selector dial.) Return throttle to 1000 rpm.

Leaving the turbo boost selector dial as set, run up engines No. 2, No. 3, and No. 4, successively to check manifold pressure. The manifold pressure of each engine should equal that set on engine No. 1. If small adjustments are needed, they can be made with the small individual potentiometers by removing the black cap and turning the screw found underneath (1) clockwise to increase manifold pressure, or (2) counter-clockwise to decrease manifold pressure.

Takeoff

1. Set turbo boost selector dial to reference point found correct in engine run-up.
2. After takeoff, re-set the turbo boost selector dial when reducing power to obtain desired manifold pressure, or to zero setting if boost is not needed. Adjust throttles to obtain desired manifold pressure for climbing.
3. Reduce rpm for climbing.
4. During the climb, continue to adjust manifold pressure with throttles until they are in the full open position. Then obtain desired manifold pressure by using the turbo boost selector dial.

Landing

During the before-landing check, set rpm and turbo boost selector dial on downwind leg, as outlined in checklist.

High Altitude

When flying at high altitude you may reach a point where further turning of the selector dial fails to produce an increase in manifold pressure. This means that the overspeed portion of the turbo governor is limiting the turbo speed to safe rpm. When you encounter this condition, turn the manifold pressure selector dial counter-clockwise until it controls manifold pressure again. This prevents undue wear of the overspeed governor mechanism.

Emergency Power

Full emergency power (war power) can be obtained at maximum engine rpm and full throttles by releasing the dial stop and turning the turbo boost selector dial up to its limit. However, this setting places heavy strain on the engines. **Use it only in emergencies and then only for periods not exceeding 2 minutes.**

USE OF TURBO-SUPERCHARGER

To save wear and tear on the turbo-supercharger and to avoid excessive carburetor air temperature, **maintain desired manifold pressure by advancing the throttles before using the turbos.** At higher altitudes, definite turbo overspeeding may result from the use of part throttle and full turbo-supercharger operation.

You will note that during the climb manifold pressure tends to increase, making it necessary to keep retarding the turbo controls to hold constant intake manifold pressure in the climb. The regulator on the B-17 does not provide a constant intake manifold pressure during the climb but it does provide constant exhaust stack pressure. Any position of the supercharger control lever corresponds to a certain exhaust manifold pressure, and the supercharger regulator unit maintains that exhaust pressure by varying the waste gate opening. Therefore, as long as the control lever is not moved, the exhaust pressure is maintained at constant value, corresponding to the position of the control lever irrespective of altitude and reduced outside temperature and pressure.

Before starting the climb, the manifold pressure is set to a certain value with the control lever. Atmospheric pressure decreases rapidly during the climb. The difference between exhaust pressure and the atmospheric pressure thus increases with altitude and results in a greater pressure differential across the turbo. This increased turbo power is transmitted to the impeller, which utilizes it to further increase the differential between atmospheric pressure and carburetor pressure. The engine internal impeller then raises the carburetor air pressure to engine manifold pressure.

By this time both the carburetor and manifold pressure have exceeded the required values, primarily because of the tremendous output Hp of the turbo at higher rpm, which more than offsets the additional power requirements of the impeller. For example, the turbo Hp developed at 30,000 feet is almost five times that of 10,000 feet altitude with constant exhaust pressure. To correct for this excess turbo Hp, the boost control lever must be pulled back during the climb in small amounts, thus reducing the exhaust pressure.

The increase in pressure differential between exhaust pressure and outside atmospheric pressure across the turbo becomes so great at high altitude that to avoid overspeed you must decrease manifold pressure 1½" for each 1000 feet above the critical altitude.

Since you have manually decreased the manifold pressure, power is reduced proportionately, but the exhaust pressure and manifold pressure are not constant for all altitudes and you must continually readjust the supercharger controls while changing altitude. At sea level, the turbo turns at only about 10,000 rpm, while at 30,000 feet it turns at 21,300 rpm, which is the recommended maximum speed for continuous operation. Thus you must reduce the manifold pressure by the amount required to keep the turbo rpm constant with increased altitude above 30,000 feet.

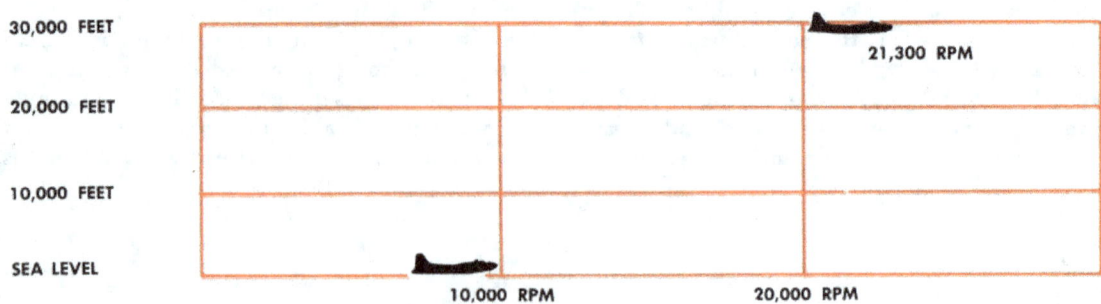

As noted, 21,300 rpm has been determined to be the maximum operating turbo speed on the type B-2 turbo, with 5% overspeed allowance in emergencies. This would provide an emergency rating of 22,400 rpm.

The pressure in the exhaust stack, and therefore the pressure just upstream from the nozzle box, depends mainly upon the amount of the exhaust gas supplied by the engine. If the engine continues to develop leading power, the exhaust stack interior continues to maintain merely the same pressure, but the pressure on the outside of the turbo wheel is atmospheric pressure which continues to decrease with increased altitude until at 30,000 feet the pressure is only about 8.9" Hg. instead of 29.92", or less than one-third that of sea level.

The velocity of the exhaust gas past the turbo buckets, and consequently the speed of the turbo wheel, depends directly on the pressure differential between the inside exhaust stack and the atmospheric pressure, or the pressure differential across the nozzle. The decrease in manifold pressure therefore must reduce the exhaust gas pressure the same amount as the lapse rate in atmospheric pressure in order to keep the nozzle pressure differential approximately constant.

Note:—For constant turbo speed at 21,300 rpm, refer to T. O. AN 01-20EF-1 for variations in manifold pressure with altitude.

Turbo Surge

You may find in certain conditions when using high turbo boost that there is a surge in manifold pressure. Turbo surge is caused by the action of the turbo-supercharger as a pressure pump. At a certain turbo rpm, the turbo will pump air into the induction system and continue to build up the induction system pressure to a certain point. Then the pressure will unload and effectively go back through the turbo against the centrifugal action of the blower. This reduces the pressure differential across the turbo and the turbo speeds up, tending to increase the induction system pressure again with its consequent reduction of turbo rpm and repetition of the cycle. This evidences itself in a surge in manifold pressure, resulting in inefficient operation and danger of temporary turbo overspeed.

The correction for this surge is to increase rpm until it discontinues. If turbo surge does not correct itself with an increase in engine rpm, the cause is probably a clogged or restricted governor balance line or a faulty governor.

Closed Turbo Waste Gate

If engine rpm is continually reduced with a wide-open throttle, manifold pressure falls off because of the closed turbo waste gate. At 25,000 feet, this begins at about 1650 rpm, while at higher altitudes it begins at higher rpm. This decrease in manifold pressure at full boost with a reduction in engine rpm takes place because as the engine rpm decreases the supply of exhaust gas from the engine is reduced. At a certain point the supply is not sufficient to drive the turbo fast enough to keep up the manifold pressure. As the rpm decreases from this point, the manifold pressure decreases also, since the turbo waste gate is closed and virtually all of the exhaust gas is going through the turbo. Reducing the exhaust gas supply reduces turbo rpm, which in turn reduces manifold pressure.

You may find that in a condition such as formation flying at high altitude, using full throttle and high boost, power often will not increase again after throttles have been retarded considerably to avoid over-running the formation. This may be especially true if the rpm has been reduced by the throttle retardation. Restoration of full throttle and increase in boost will not bring up the manifold pressure.

This often results from the fact that the power may have been reduced to the region of closed waste gate, and insufficient exhaust gas is available to turn the turbo fast enough to bring up the manifold pressure. The correct procedure under this condition is to increase the rpm to 2500, if necessary, and keep the boost setting high until manifold pressure comes up. It usually comes up immediately.

HEATING AND VENTILATING SYSTEM

The B-17F airplane has a main and an auxiliary heating system, both of which operate on the same principle of heat exchange.

Main System

The main system supplies cabin heat through a glycol system in nacelle No. 2.

The heating system fluid (glycol solution of 55% diethylene glycol and 45% ethylene glycol by weight) is stored in a tank in the top of nacelle No. 2. The glycol flows from the tank to the engine-driven pump, which circulates the fluid at a rate of 55 to 60 U.S. gallons per hour. The flow is directed to a filter which removes impurities from the fluid. The glycol is then pumped through 3 heaters, which are installed in series and located in the exhaust stack, where it collects the heat of the exhaust gases.

A relief valve, between pump and filter, bypasses the glycol back to the supply line if high pressure is built up in the system during cold weather, or if the heaters are clogged.

The circulation of the glycol is continuous and therefore it must be constantly cooled. For this purpose there is a radiator between the spars in the left-hand wing gap. Ram air from the intercooler air inlet absorbs heat from the glycol at the radiator, and passes through the radiator and into the cabin. The cooled glycol passes into the supply tank. A controllable damper in the radiator may be operated to spill the air overboard if desired.

Auxiliary System (Some B-17F's)

The auxiliary heating system uses the same principle of heat exchange as that employed by the normal heating system and has a heater unit, filter, relief valve, pump and supply tank installation in nacelle No. 3 identical to the corresponding installation of the main heating system in nacelle No. 2. Eight radiator-fan assemblies are connected by glycol tubing to the heater units in nacelle No. 3. Five of these are the non-recirculating type (external radiator air supply) and the remaining are the recirculating type (internal radiator air supply). The non-recirculating type radiator-fan assemblies are in the astrodome, top turret, ball turret, and the tail gun enclosure. Each of these assemblies has a hand-operated damper which directs the flow of heated air to the gun and/or windows, or spills it overboard. The recirculating radiator-fan assemblies have overboard discharge ducts and damper tube controls for regulating the amount of heated air admitted to the pilot's, navigator's, and radio operator's compartment. Electric fan control is automatic.

Two thermo-switches, mounted on the glycol tubing under the flow of the pilots' compartment, turn 5 non-recirculating radiator fans on at 177°C (350°F) and the 3 recirculating radiator fans on at 77°C (150°F). The thermo-switches are capable of functioning when the master switch is "ON."

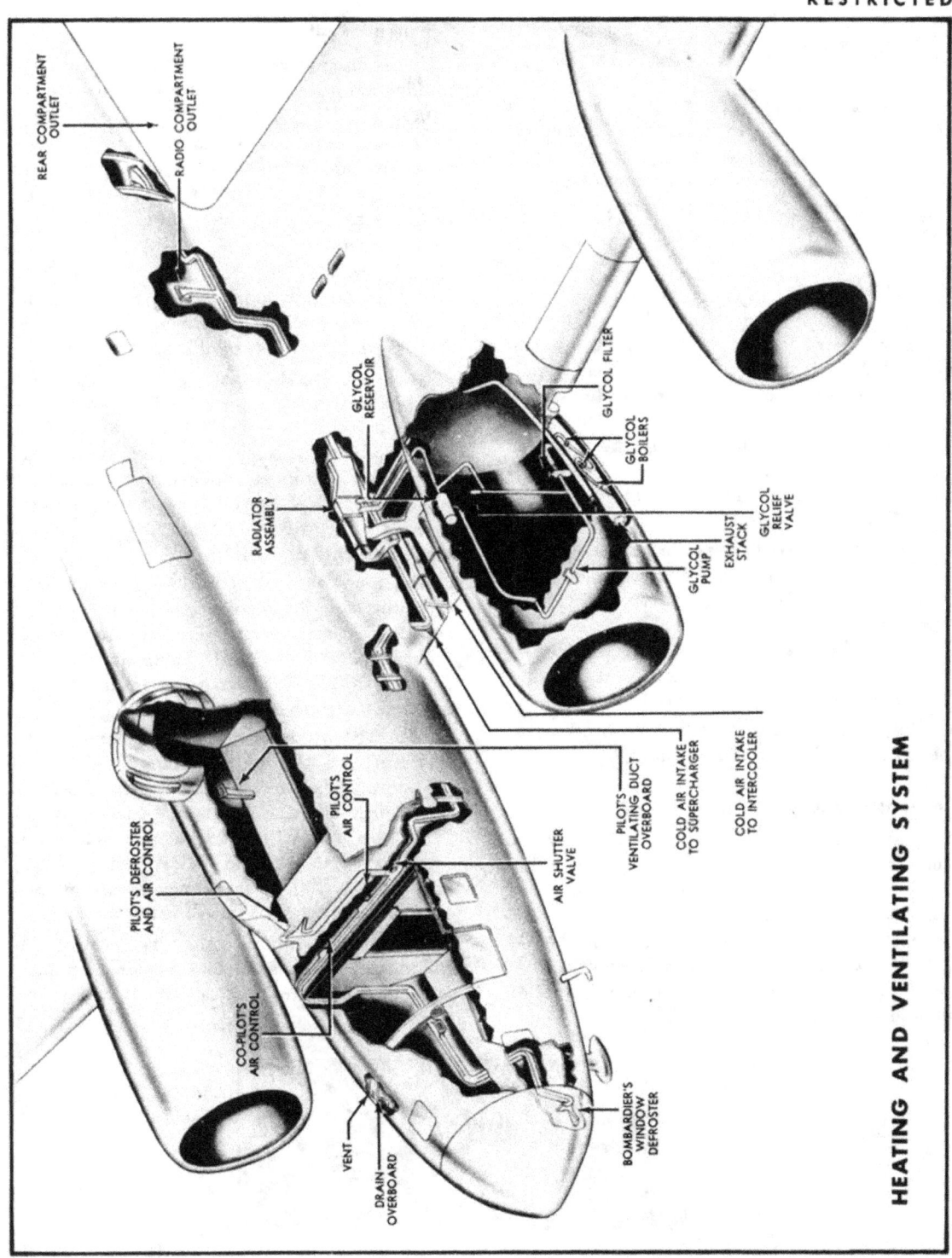

HEATING AND VENTILATING SYSTEM

RESTRICTED

COMMUNICATION

The B-17 contains equipment for long and short-range two-way voice and code communication, intercommunication between crew members, emergency transmission, directional indication, and reception of marker beacon signals.

Interphone

The interphone system provides for communication between crew members. Command radio, liaison radio, and radio compass signals are audible over the interphone system at all crew stations. Any crew station can talk over the command transmitters. Only the pilot, copilot, navigator, and radio operator can transmit over the liaison radio.

Interphone equipment includes a dynamotor and amplifier located under the radio operator's table, and 12 jackboxes located throughout the airplane: 3 in the nose (for the navigator, bombardier, and forward gunner), 3 on the flight deck (for the pilot, copilot, and top turret gunner), 2 in the radio compartment, 3 in the waist compartment, and one in the tail compartment.

Remember: Crew members should wear headsets at all times during flight.

Interphone Call

The "CALL" position on the jackbox enables the user to over-ride reception on all other jackbox stations for the purpose of calling any particular station. A spring returns the selector switch to "INTER" so that it cannot be left in the "CALL" position inadvertently. For obvious reasons, use of the "CALL" position should be held to a minimum.

Command Radio

The command radio is for short-range communication with aircraft and ground stations.

Voice transmission over the command set is available to all crew stations, but code transmission is limited to the pilot and copilot, who alone have a transmitting key. It is on the remote control box on the ceiling of the pilot's compartment.

The command radio consists of 3 receivers and 2 transmitters on the right forward bulkhead of the radio compartment. Remote controls are on ceiling of pilot's compartment.

Remote Control Units: The transmitter control box has an on-off toggle switch which turns on either transmitter, and a transmitter selector switch which selects either of the 2 transmitters. (Positions are provided for 4 transmitters, should the 2 extras be installed.) A wave selector switch turns on voice, CW (continuous wave) or tone as desired.

The receiver control is divided into 3 control units, one for each receiver. The low band receiver covers 190-550 Kc, the intermediate band from 3000 to 6000 Kc, and the high band from 6000-9100 Kc. Each receiver control unit has 2 switches to operate it.

The A-B switch selects either jackbox or control unit. Use "A" if plugged into jackbox; use "B" if plugged directly into control unit. **A tone selector switch** which can select "TONE," "CW," or "MCW" should be turned to modulated CW with "A" and "MCW" on. Then you can tune to desired frequency by means of a small handle which turns a calibrated dial.

The reliable transmitting range of the command set is 25 miles or less. Under good atmospheric conditions greater range may be obtained.

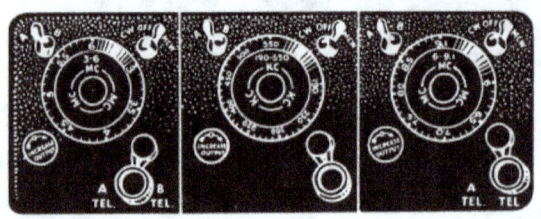

176   RESTRICTED

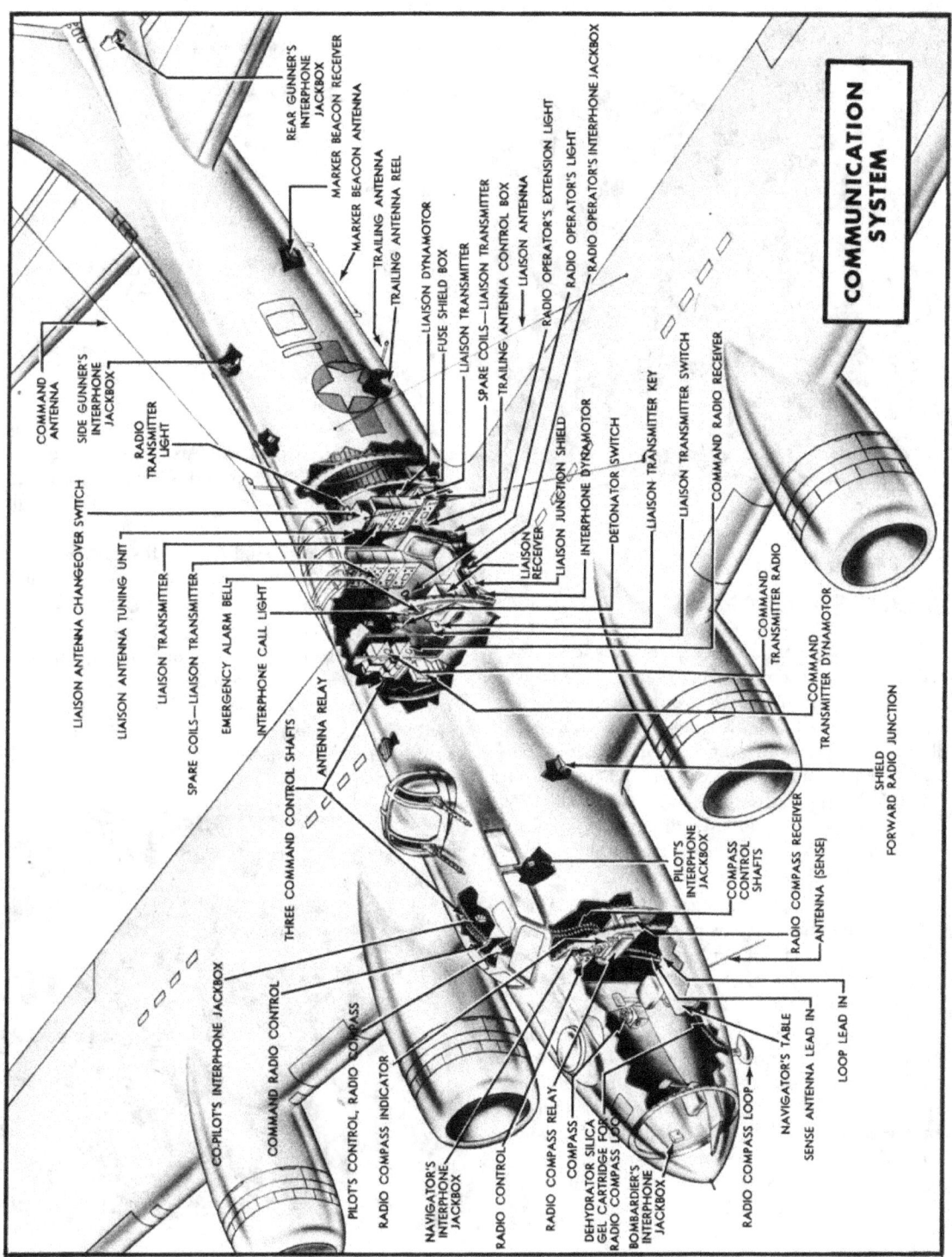

LIAISON TRANSMITTER

This transmitter, on the aft bulkhead of the radio compartment, insures communication with aircraft in flight and ground stations over distances up to 3000 miles, depending on atmospheric conditions and method of transmission. The usual reliable distances are 250 miles on voice, 500 miles on tone and 750 miles on CW. Only 4 jackbox positions (radio operator, pilot, copilot and navigator) can transmit on the liaison set.

This set has 7 interchangeable turning units covering frequencies from 360-650 Kc and 1500-12,500 Kc, and including a low band from 200-500 Kc in some models. For tuning this set, see communication section of B-17 T.O.'s.

The liaison receiver on the radio operator's table covers a frequency range from 1500-18,000 Kc. It uses the same antenna as the transmitter: the skin of the airplane. This is connected to a throw switch on the left side wall. This switch can change over to the trailing antenna (also on left side wall). The trailing antenna is operated from a control box to the right of the change-over switch.

RADIO COMPASS (SCR 269-G)

The radio compass is a multi-purpose receiver designed primarily as a navigational instrument.

The power for this set comes from the airplane's batteries and inverters. The various relays and switches operate on the direct current supply, and the receiver and motors for rotation of the loop operate on the inverters.

This set has 2 antennas: a sensing (whip), or non-directional antenna, and a loop, or directional antenna.

The radio compass is a multi-band receiver and, as installed in B-17 aircraft, may be remotely controlled from either of 2 identical control boxes. One of these boxes is above and between the pilot and copilot; the other directly above and slightly to the left of the receiver itself in the navigator's compartment.

Each of the control boxes (BC-434-A) has (1) an antenna selector switch; (2) a band-change switch; (3) a control button; (4) a main tuning control; (5) a tuning indicator (meter); (6) an audio output control; (7) a loop L-R switch; (8) a control light; (9) a dial light, and (10) a dial light control.

Antenna Selector Switch

This four-position ("OFF - COMP - ANT - LOOP") switch selects the type antenna or antennas to be used, AVC or MVC, and largely determines the indication and action of the loop. The 4 positions of this switch may be explained as follows:

"OFF": Self-explanatory.

"COMP": When in this position, the set is using both sensing (whip) and loop antennas. Automatic volume control is always present in this position, and the operation of the radio compass indicators and loop is automatic.

"ANT": This position utilizes only the whip or non-directional antenna; therefore the loop and indicators do not operate. Manual volume control is now present, and the volume is adjusted or regulated only by means of the audio control. This position should be used at all times for the initial tune-in of the station.

"LOOP": Now only the loop or directional antenna is in use. The operation of the loop and indicators are controlled by means of the Loop L-R switch. Again you have only manual volume control.

Band-change Switch

This electrically controlled switch selects the band or range of frequencies desired. There are 3 positions, or bands. One band covers frequencies from 200 to 410 Kc; another from 410 to 850 Kc; the third from 850 to 1750 Kc.

Tuning Control

After selecting the desired band with the band change switch, use this control to select any desired frequency within this band.

Control Light

This light on the control box indicates the control box actually controlling the compass receiver. When the light is on, the control box is in control of the radio compass.

Control Button

This push button throws control of the radio compass from one control box to the other. If the light on the desired control box is unlighted, press this control. When you release it, control is switched from the other control box to the desired box.

Audio

By varying this control, the operator may adjust the headset volume as desired.

Loop L-R

This switch controls rotation of the loop when the antenna switch is in "LOOP" position. The loop can be rotated at two different speeds. When the Loop L-R switch is pressed in and switched to the desired position (L rotates loop to left, R to right) and held there, the loop rotates at a fairly rapid speed. When the switch is not pushed in, but only held in the desired position, the loop rotates slowly. When the loop is rotated by this switch the compass indicators rotate to show the position of the loop.

Tune for Maximum Indication

When tuning in any station, the main tuning control should be tuned for maximum swing of the needle on "Tune for Maximum" indicator.

Dial Light Control

This control regulates the brilliance of the dial light.

For instructions on how to use the radio compass, see **Advanced Instrument Flying**, T.O.–30-100.

RADIO SET SCR 522 A

The SCR 522 A VHF (very high frequency) transmitter-receiver radio set provides 2-way radio-telephone communication between aircraft in flight and between aircraft and ground stations. Provision is made for voice communication and continuous audio-tone modulation.

The pilot and co-pilot control the SCR 522 by means of the radio control box on the left side of the pilot's control pedestal in the B-17. The set operates on any one of 4 pre-set crystal-controlled frequency channels lying within the range of 100-156 Mc. Line-of-sight communication is normally necessary for satisfactory operation of the radio set.

The following table lists the approximate range to be expected, assuming communication is taking place between the aircraft and a ground station over level country.

Altitude above ground	Approximate range
1000 feet	30 miles
3000	70
5000	80
10,000	120
15,000	150
20,000	180

Radio Control Box

The radio control box to the left of the pilot's control pedestal provides the only complete remote control of communications functions. Five red push buttons are the means by which any one of the 4 channels (A, B, C, and D) is selected and the power turned off. When the "OFF" push button is pressed, the dynamotor is stopped. The push buttons are interconnected so that not more than one channel can be selected at a given time. A light opposite each push button indicates which channel is being used.

The "T-R-REM" switch (transmit-receive-remote) is normally in the "REM" position, permitting press-to-talk operation by means of the conventional push button microphone switch on the pilot's control wheel, which when depressed switches the equipment from **receive** to **transmit**. In the "T" position the transmitter is in continuous operation. In the "R" position the receiver is in continuous operation.

The lever tab, directly above the "T-R-REM" switch, when lowered, blocks the switch from "REM" position and spring-loads the switch lever so that unless it is held in the "T" position it will return to "R."

The small lever tab opposite the "OFF" push button is a dimmer mask to reduce the lamp glare. The lamp opposite the "T-R-REM" switch is on when receiving and off when transmitting.

Transmitter-Receiver Assembly

The transmitter and receiver units are in a single case. The transmitter employs a crystal-controlled oscillator circuit and operates in the frequency range of 100-156 Mc on one of the 4 pre-set channels A, B, C, and D. Average output power of the transmitter is 8 to 9 watts, using a total power input current of 11.5 amps at 28 volts.

The receiver is a sensitive superheterodyne unit employing a heterodyne oscillator whose frequency is controlled by any one of 4 quartz crystals. Thus the 4 crystal-controlled channel frequencies within the range 100-156 Mc are available for instantaneous selection at the remote control position. For reception the total input current is 11.1 amps at 28 volts.

Dynamotor Unit

The dynamotor operates on the 28-volt power circuit and supplies 3 regulated voltage sources (300-volt DC, 150-volt DC, and 13-volt DC) required for operation of the transmitter-receiver assembly.

In addition to the equipment listed above, jackboxes, junction boxes, headsets, and microphones are used with the radio set.

Operation of the SCR 522 A

1. **Transmission only**

To start the equipment, press push button A, B, C, or D depending upon which channel is to be used.

Allow approximately one minute for the vacuum tubes to warm up.

Move the "T-R-REM" switch to the "T" position and speak into the microphone.

2. **Reception only**

Place the "T-R-REM" switch in the "R" position. It is held in the "R" position by lowering the small lever tab.

To start the equipment press push button A, B, C, or D for the desired channel.

3. **Press-to-transmit (press-to-talk) operation**

Place the "T-R-REM" switch in the "REM" position.

To start the equipment select a channel by pressing push button A, B, C, or D.

To receive: Under these conditions the receiver is normally in continuous operation.

To transmit: Depress the press-to-talk button and speak into the microphone.

To receive again: Release the press-to-talk microphone button.

4. **To shut off the equipment**, press the "OFF" button.

Common Uses of Channels

"A" channel is usually used for all normal plane-to-plane communication or for plane-to-ground communication with a Controller.

"B" channel is common to all VHF-equipped control towers. It is normally used to contact the control tower for takeoff and landing instructions.

"C" channel is frequently used in contacting homing stations.

"D" channel is normally used for plane-to-ground contact with D/F stations, and as a special frequency which is automatically selected at regular intervals by the action of a contactor unit.

Precautions During Operation

Avoid prolonged use of the radio on the ground to conserve the batteries and avoid overheating of the dynamotor.

If the transmitter and receiver fail to operate when a channel push button is pressed on the radio control box, press another channel push button, then again press the push button for the desired channel. Transmission and reception should now be possible.

FREQUENCY METER

A frequency meter is standard equipment on all B-17's and should be kept in the radio compartment. It is used to check and correct transmitters and receivers on frequency ranges from 125 Kc to 20,000 Kc. For use and operation, see Technical Orders.

MARKER BEACON

The radio marker beacon receiver receives ultra-high frequency signals used in aircraft navigation and landing, and reproduces them visually by an amber light on the pilot's instrument panel. When the receiver is over a keyed transmitter, such as a C.A.A. marker, or certain types of Army transmitters, the indicator lamp flashes in accordance with the identifying signal of the transmitter.

EMERGENCY OPERATION OF RADIO EQUIPMENT

Interphone Equipment Failure

If the interphone equipment fails, the audio frequency section of the command transmitter may be substituted for the regular interphone amplifier. To make this connection, the pilot places his command transmitter control box channel selector switch in either channel No. 3 or 4 position (or whatever position is not being used with a transmitter). Set the interphone jackbox selector switch to "COMMAND" to place the interphone equipment in operation.

When the command transmitter control box channel selector switch is set in either the No. 3 or 4 position for emergency operation of the interphone equipment, it is not possible to establish communication with any ground station or any other airplane. It is possible at all times to resume normal command set operation by placing the channel selector switch of the command transmitter control box in either the No. 1 or 2 position.

Substitution of Radio Compass Receiver for Low-Frequency Command Set Receiver

If the low-frequency receiver of the command set fails, the radio compass receiver may be substituted, with the pilot having **direct control** over the compass receiver. To complete this emergency hookup, the pilot must set his interphone jackbox selector switch in the "COMP" position and then place the radio compass selector switch in the "ANT" position. The radio compass can then be tuned as desired.

Substitution of Liaison Receiver for Low, Medium, and/or High-Frequency Command Receiver

In case of the failure of the low, medium, and/or high-frequency receiver of the command radio equipment, the liaison receiver may be substituted, but the pilot will have only limited control over it. The pilot should first call the radio operator on the interphone system and tell him what frequency he desires to receive, that he is switching the interphone selector switch to the "LIAISON" position, and for him (the radio operator) to tune in this frequency and maintain the setting until further notice.

Command Set Transmitter Failure

If the command set transmitter fails, the liaison transmitter may be substituted. The pilot should first call the radio operator on the interphone and have him adjust the liaison transmitter to the frequency he desires to use. He should then set his interphone selector switch to the "LIAISON" position and operate his microphone button in the same manner that he did when the command set was in operation. When he is through using the liaison transmitter, the pilot should place the interphone selector switch in the "INTER" position and tell the radio operator to cut the liaison transmitter off, to reduce the load on the electrical system.

When substituting one receiver for another, such as the compass receiver for the command receiver, the pilot must move his interphone selector switch to the "COMMAND" or "LIAISON" position, as the case may be, in order to transmit. At the end of the transmission, he must switch back to the position of the receiver being used. He must do this every time he desires to hold a 2-way conversation.

THE C-1 AUTO PILOT

The C-1 autopilot is an electromechanical robot which automatically controls the airplane in straight and level flight, or maneuvers the airplane in response to the fingertip control of the human pilot or bombardier.

Actually, the autopilot works in much the same way as the human pilot in maintaining straight and level flight, in making corrections necessary to hold a given course and altitude, and in applying the necessary pressure on the controls to turns, banks, etc. The difference is that the autopilot acts instantaneously and with a precision that is not humanly possible.

The precision of even the most skillful human pilot is limited by his own reaction time, i.e., the interval between his perception of a certain condition and his action to correct or control it. Reaction time itself is governed by such human fallibilities as fatigue, inability to detect errors the instant they occur, errors in judgment and muscle coordination.

The autopilot, on the other hand, detects flight deviations the instant they occur, and just as instantaneously operates the controls to correct the deviations. Properly adjusted, the autopilot will neither overcontrol nor undercontrol the airplane, but will keep it flying straight and level with all 3 control surfaces operating in full coordination.

How It Works

The C-1 autopilot consists of various separate units electrically interconnected to operate as a system. The operation of these units is explained in detail in AN-11-60AA-1. A general over-all understanding of their functions and relation to each other can be acquired by studying the accompanying illustration.

Assume that the airplane in the illustration is flying straight and level and that the autopilot is at work.

Suddenly rough air turns the airplane away from its established heading. The gyro-operated* directional stabilizer (1) in the bombardier's compartment detects this deviation and moves the directional panel (4) to one side or the other, depending upon the direction of the deviation.

RESTRICTED

The directional panel contains 2 electrical devices, the banking pot (5) and the rudder pick-up pot (6), which send signals to the aileron and rudder section of the amplifier (16) whenever the directional panel is operated. These signals are amplified and converted (by means of magnetic switches or relays) into electrical impulses which cause the aileron and rudder Servo units (15 and 18) to operate the ailerons and rudder of the airplane in the proper direction and amount to turn the airplane back to its original heading.

Similarly, if the nose of the airplane drops, the vertical flight gyro (10) detects the vertical deviation and operates the elevator pick-up pot (11) which sends an electrical signal to the elevator section of the amplifier. The signal is amplified and relayed in the form of electrical impulses to the elevator Servo unit (19) which in turn raises the elevators the proper amount to bring the airplane to level flight.

If one wing drops appreciably, the vertical flight gyro operates the aileron pick-up pot (12), the skid pot (13), and the up-elevator pot (14). The signals caused by the operation of these units are transmitted to their respective (aileron, rudder, and elevator) sections of the amplifier. The resulting impulses to the

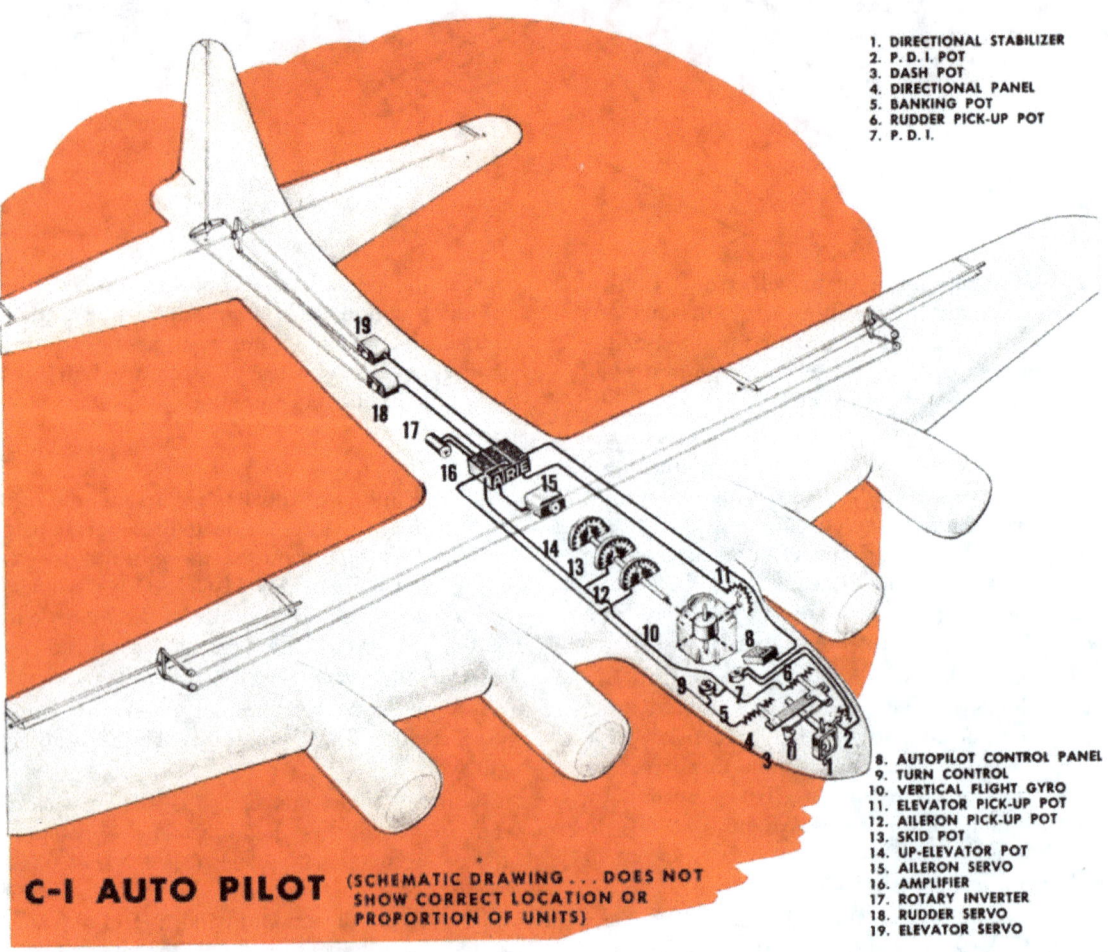

1. DIRECTIONAL STABILIZER
2. P. D. I. POT
3. DASH POT
4. DIRECTIONAL PANEL
5. BANKING POT
6. RUDDER PICK-UP POT
7. P. D. I.

8. AUTOPILOT CONTROL PANEL
9. TURN CONTROL
10. VERTICAL FLIGHT GYRO
11. ELEVATOR PICK-UP POT
12. AILERON PICK-UP POT
13. SKID POT
14. UP-ELEVATOR POT
15. AILERON SERVO
16. AMPLIFIER
17. ROTARY INVERTER
18. RUDDER SERVO
19. ELEVATOR SERVO

C-1 AUTO PILOT (SCHEMATIC DRAWING... DOES NOT SHOW CORRECT LOCATION OR PROPORTION OF UNITS)

aileron, rudder, and elevator Servo units cause each of these units to operate its respective control surface just enough to bank and turn the airplane back to a level-flight attitude.

When the human pilot wishes to make a turn, he merely sets the turn control knob (9) at the degree of bank and in the direction of turn desired. This control sends signals, through the aileron and rudder sections of the amplifier, to the aileron and rudder Servo units which operate ailerons and rudder in the proper manner to execute a perfectly coordinated (non-slipping, non-skidding) turn. As the airplane banks, the vertical flight gyro operates the aileron, skid, and up-elevator pots (12, 13, 14). The resulting signals from the aileron and skid pots cancel the signals to the aileron and rudder Servo units to streamline these controls during the turn.

The signals from the up-elevator pot cause the elevators to rise just enough to maintain altitude. When the desired turn is completed, the pilot turns the turn control back to zero and the airplane levels off on its new course. A switch in the turn control energizes the directional arm lock on the stabilizer, which prevents the stabilizer from interfering with the turn by performing its normal direction-correcting function.

The autopilot control panel (8) provides the pilot with fingertip controls by which he can conveniently engage or disengage the system, adjust the alertness or speed of its responses to flight deviations, or trim the system for varying load and flight conditions.

The pilot direction indicator, or PDI (7), is a remote indicating device operated by the PDI pot (2). When the autopilot is used, the PDI indicates to the pilot when the system and airplane are properly trimmed. Once the autopilot is engaged, with PDI centered, the autopilot makes the corrections automatically.

The rotary inverter (17) is a motor-generator unit which converts direct current from the airplane's battery into 105-cycle alternating current for operation of the autopilot.

HOW TO OPERATE THE C-1 AUTOPILOT

Before Takeoff

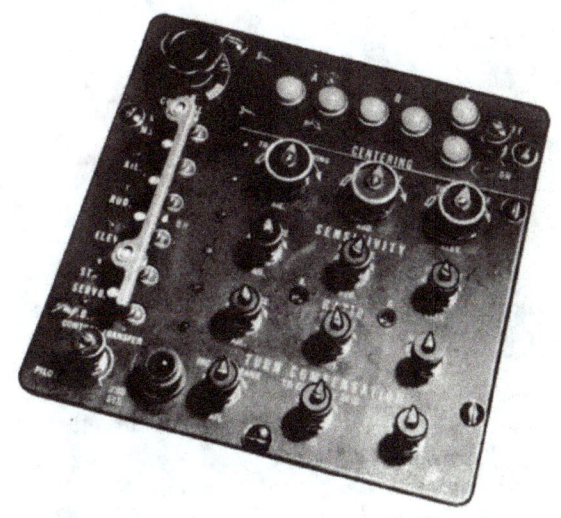

1. Set all pointers on the control panel in the up position.

2. Make sure that all switches on the control panel are in the "OFF" position.

After Takeoff

1. Turn on the master switch.

 SERVO PDI

2. Five minutes later, turn on PDI switch (and Servo switch, if separate).

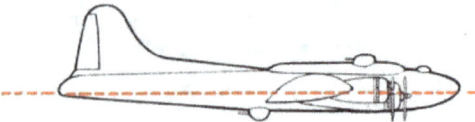

3. Ten minutes after turning on the master switch, trim the airplane for level flight at cruising speed by reference to flight instruments.

4. Have the bombardier disengage the autopilot clutch, center PDI and lock it in place by depressing the directional control lock. The PDI is held centered until the pilot has completed the engaging procedure. Then the autopilot clutch is re-engaged, and the directional arm lock released.

Alternate Method: The pilot centers PDI by turning the airplane in direction of the PDI needle. Then resume straight and level flight.

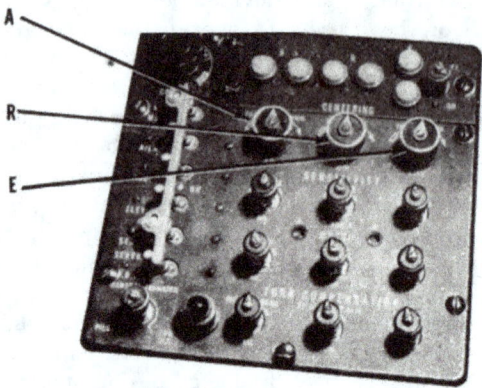

5. Engage the autopilot. Put out aileron telltale lights with the aileron centering knob, then throw on the aileron engaging switch. Repeat the operation for rudder, then for elevator.

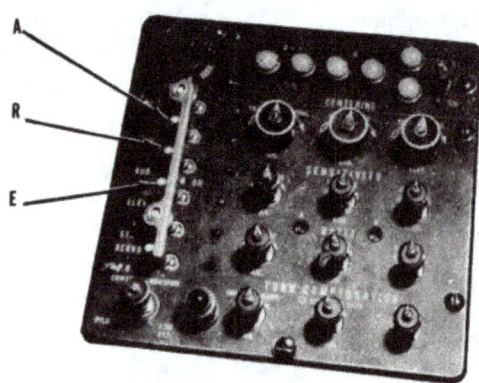

6. Make final autopilot trim corrections. If necessary, use centering knobs to level wings and center PDI.

Caution:

NEVER ADJUST MECHANICAL TRIM TABS WHILE THE AUTOPILOT IS ENGAGED

FLIGHT ADJUSTMENTS AND OPERATION

After the C-1 autopilot is in operation, carefully analyze the action of the airplane to make sure all adjustments have been properly made for smooth, accurate flight control.

When both **tell-tale** lights in any axis are extinguished, it is an indication the autopilot is ready for engaging in that axis.

Before engaging, each **centering knob** is used to adjust the autopilot control reference point to the straight and level flight position of the corresponding control surface. After engaging, centering knobs are used to make small attitude adjustments.

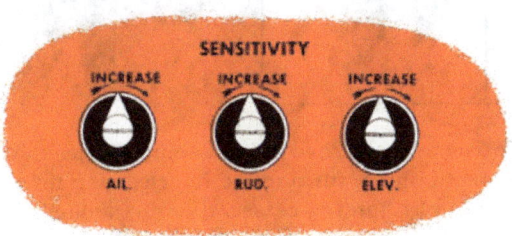

Sensitivity is comparable to a human pilot's reaction time. With sensitivity set high, the autopilot responds quickly to apply a correction for even the slightest deviation. If sensitivity is set low, flight deviations must be relatively large before the autopilot will apply its corrective action.

Ratio is the amount of control surface movement applied by the autopilot in correcting a given deviation. It governs the speed of the airplane's response to corrective autopilot actions. Proper ratio adjustment depends on airspeed.

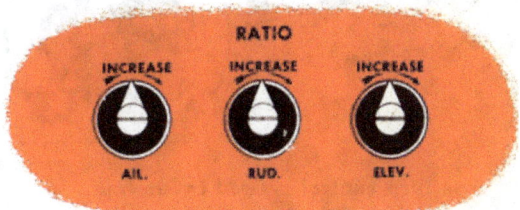

If ratio is too high, the autopilot will overcontrol the airplane and produce a ship hunt; if ratio is to low, the autopilot will undercontrol and flight corrections will be too slow. After ratio adjustments have been made, centering may require readjustment.

To adjust **turn compensation**, have bombardier disengage autopilot clutch and move engaging knob to extreme right or extreme left. Airplane should bank 18° as indicated by artificial horizon. If it does not, adjust aileron compensation (bank trimmer) to attain 18° bank. Then, if turn is not coordinated, adjust rudder compensation (skid trimmer) to center inclinometer ball. Do not use aileron or rudder compensation knobs to adjust coordination of turn control turns.

Emergency Use of Autopilot

REMEMBER THE ROLE THAT THE AUTOPILOT CAN PLAY IN EMERGENCIES

1. If the control cables are damaged or severed between the pilot's compartment and the Servo units in the tail, the autopilot can bridge the gap. There have been many instances where the autopilot has been used thus to fly an airplane with damaged controls.

2. If the autopilot has been set up for level flight, it can be used to hold the airplane straight and level while abandoning ship.

RESTRICTED

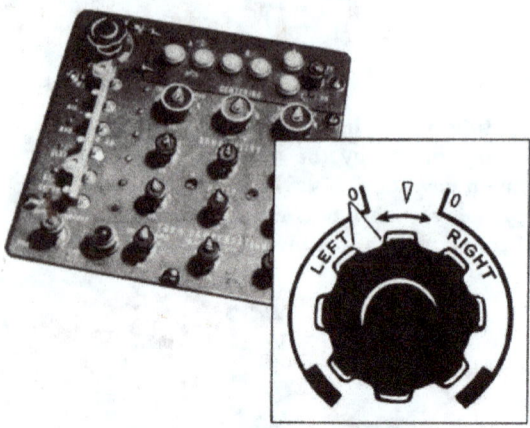

The **turn control** is used by the pilot to turn the airplane while flying under automatic control. To adjust turn control, first make sure turn compensation adjustments have been properly made, then set turn control pointer at beginning of trip-lined area on dial. Airplane

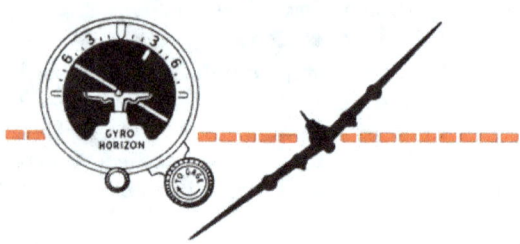

should bank 30°, as indicated by artificial horizon. If it doesn't, remove cap from aileron trimmer and adjust trimmer until a 30° bank is attained. Then, if turn is not coordinated (inclinometer ball not centered), adjust rudder trimmer to center ball. Make final adjustments with both trimmers and replace caps. Set turn control at zero to resume straight and level flight; then re-center.

Never operate the Turn Control without first making sure the PDI is centered

The **turn control transfer** has no effect unless the installation includes a remote turn control.

The **dashpot** on the stabilizer regulates the amount of rudder kick applied by the autopilot to correct rapid deviations in the turn axis. If a rudder hunt develops which cannot be eliminated by adjustment of rudder ratio or sensitivity, the dashpot may require adjustment.

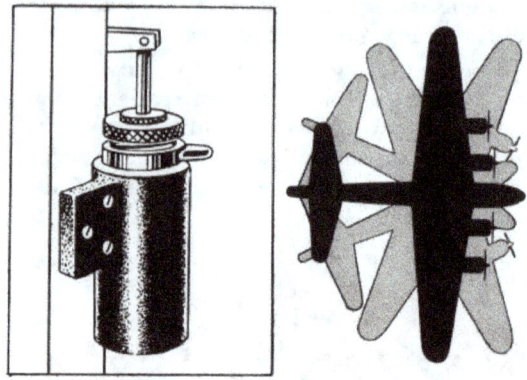

This is accomplished by loosening the locknut on the dashpot, turning the knurled ring up or down until hunting ceases, then tightening the locknut.

Cold Weather Operation—When temperatures are between —12° and 0°C (10° and 32°F) autopilot units must be run for 30 minutes before engaging. If accurate flight control is desired immediately after takeoff, perform the autopilot warm-up before takeoff by turning on the master switch during the engine run-up—but make sure autopilot is off during takeoff. If warm-up is performed during flight, allow 30 minutes after turning on master switch before engaging. When temperatures are below —12°C (10°F) units must be preheated for one hour before takeoff. Use special heating covers or blankets with heating tubes.

RESTRICTED

FLYING THE PDI MANUALLY

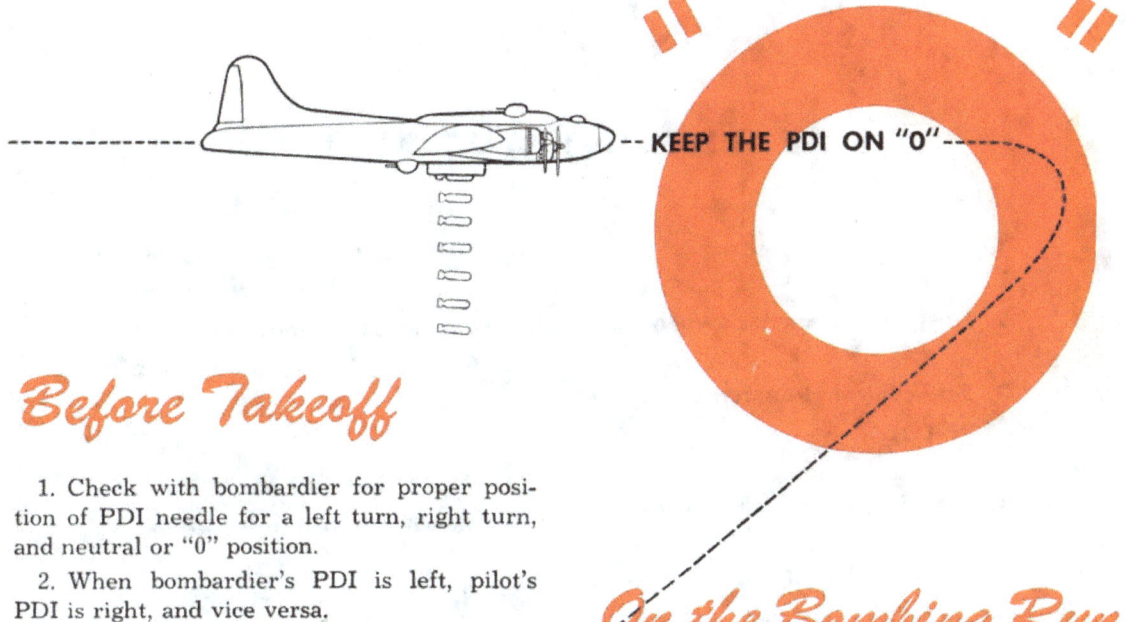

Before Takeoff

1. Check with bombardier for proper position of PDI needle for a left turn, right turn, and neutral or "0" position.
2. When bombardier's PDI is left, pilot's PDI is right, and vice versa.

On the Bombing Run

Normally bombing will be done while using the autopilot. However, if the autopilot is not functioning the pilot may use the PDI.

1. To center the PDI needle, turn the airplane in the direction of the needle.
2. At the beginning of the bombing run, the pilot usually can expect maximum PDI corrections. Avoid tendency to overcorrect by refraining from leading the needle.
3. No matter how slight the deviation of the PDI needle from "0," the needle must be returned to "0" immediately.
4. Set turns must be coordinated aileron and rudder turns, in order to make the desired degree of turn more rapidly and to avoid any excessive sliding of the bombsight lateral bubble and induced precession of the gyro.
5. To avoid tumbling of the bombsight gyro, banks must never exceed 18°.
6. Keep PDI on "0" until bombardier calls "Bombs away."

Pilot's Ground Checklist

FOR THE C-1 AUTOPILOT

1. Center turn control.

2. Turn on C-1 master switch bar.

3. Set control transfer knob at "PILOT."

4. Set tell-tale light shutter switch "ON."

5. Set all adjustment knobs to pointers-up position, making sure pointers are not loose.

6. Tell bombardier to center PDI.

7. Turn on Servo PDI switch.

8. Operate controls through extreme range several times, observing that tell-tale lights flicker and go out as streamline position is reached from either direction.

9. Turn on aileron, rudder, and elevator switches.

10. Turn aileron centering knob clockwise, then counter-clockwise, observing that wheel turns to the right and then to the left.

11. Repeat Item 10 for rudder and elevator, observing action.

12. Have bombardier move directional arm for full right turn, then to left, observing to see if aileron and rudder move in proper direction.

13. Have bombardier center PDI and engage secondary clutch.

14. Rotate turn control knob for right and left turns, observing aileron and rudder controls for proper movement.

15. If all checks are satisfactory, turn the C-1 master switch bar "OFF."

THE GYRO FLUX GATE COMPASS

The gyro flux gate compass, remotely located in the wing or tail of the airplane, converts the earth's magnetic forces into electrical impulses to produce precise directional readings that can be duplicated on instruments at all desired points in the airplane.

Unlike the magnetic needle, it will not go off its reading in a dive, overshoot in a turn, hang in rough weather, or go haywire in polar regions.

Development of the Flux Gate

The gyro flux gate compass was developed to fill the need for an accurate compass for long-range navigation. The presence of so many magnetic materials (armor, electrical circuits, etc.) in the navigator's compartment made it almost impossible to find a desirable location for the direct-reading magnetic compass.

To eliminate this difficulty, it became necessary to place the magnetic element of the navigator's compass outside the compartment, **i.e., to use a remote indicating compass.** The unit which is remotely located is called the **transmitter.** The unit used by the navigator is the **master indicator.** For the benefit of the pilot and such other crew members as may have needs for compass readings, auxiliary instruments called **repeater indicators** may be installed in other parts of the airplane.

Units of the Flux Gate Compass

The gyro flux gate compass consists of 3 units which are analogous to the brain, heart, and muscles of the human body. The **transmitter,** located in the wing or tail of the airplane, is the brain of the instrument. The **amplifier** is the source of power for the compass and corresponds to the human heart. The **master indicator** does the work of turning a pointer and performs a function similar to that of the muscles in the human body.

1. **The Brain.**—Inside the remotely placed **transmitter** there is a magnetic sensitive element called the flux gate which picks up the

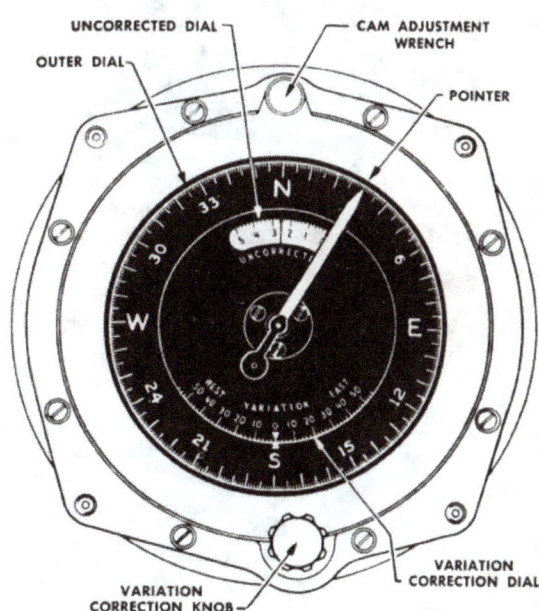

direction signal by induction and transmits it to the **master indicator.** This element consists of 3 small coils, arranged in a triangle and held on a horizontal plane by a gyro. Each coil has a special soft iron core, and consists of a primary (or excitation) winding, and a secondary winding from which the signal is obtained.

Because each leg of the flux gate is at a different angle to the earth's magnetic field, and the induced voltage is relative to the angle, each leg produces a different voltage. When the angular relationship between the flux gate and the earth's magnetic field is changed, there is a relative change in the voltages in the 3 legs of the secondary. These voltages are the motivating force for the gyro flux gate compass master indicator which provides indications of the exact position of the flux gate in relation to the earth's magnetic field.

Each coil is a direction sensitive element; but one alone would provide an ambiguous reading because it could tell north from east, for instance, but not north from south. Therefore, it

is necessary to employ 3 coils and combine their output to give the direction signal.

2. **The Heart.**—The amplifier furnishes the various excitation voltages at the proper frequency to the transmitter and master indicator. If amplifies the autosyn signal which controls the master indicator and serves as a junction box for the whole compass system.

Power for the amplifier comes from the airplane's inverter and is converted to usable forms for other units. The input of the amplifier is 400-cycle alternating current and various voltages may be used depending upon the source available.

3. **The Muscle.**—The master indicator is the muscle of the system because it furnishes the mechanical power to drive the pointer on the main instrument dial. The pointer is driven through a cam mechanism which automatically corrects the reading for compass deviation so that a corrected indication is obtained on all headings. The shaft of the pointer is geared to another small transmitting unit in the master indicator which will operate as many as six repeat indicators at other locations.

The amplifier, master indicators and repeaters all are unaffected by local magnetic disturbances.

How to Operate the Compass

1. Leave the toggle switch on the flux gate amplifier "ON" at all times so that the compass will start as soon as the airplane's inverter is turned on.

2. Leave the caging switch in the "UNCAGE" position at all times except when running through the caging cycle.

3. About 5 minutes after starting engines, throw caging switch to "CAGE" position. Leave it there about 30 seconds and then throw to "UNCAGE" again.

4. With the new push button-type caging switch, depress it for a few seconds until a red signal light goes on. Then release the switch and the caging cycle is automatically completed, at which time the red light goes out.

5. Set in the local variation on the master indicator if you wish the pointer to read true heading.

6. If at any time during flight the compass indications lead you to suspect that the gyro is off vertical, run through the caging cycle when the airplane is in normal flight attitude, especially when leveling off after climb.

Note: For further details concerning functions, operation and flight instructions, see Technical Order No. 05-15-27.

EMERGENCY EQUIPMENT

Engine Section Fire Extinguisher

1. Some planes have a remotely controlled fire extinguisher equipment to permit the copilot to discharge CO_2 into the engine accessory compartment. A selector valve for directing the CO_2 to any one of the 4 engines, and 2 pull handles, are on the auxiliary control panel in front of the copilot.

2. Two 7¼-lb. CO_2 cylinders are installed in a gap in the right wing just forward of the rear spar. Control for release of CO_2 from the cylinders is accomplished individually for each cylinder by means of flexible cables extending from the 2 pull handles in the cockpit.

3. To operate, the copilot sets the selector valve to the engine on fire and pulls the handle. Pull the other handle if the second charge of CO_2 is necessary.

Fire Extinguishers

1. There are 3 carbon dioxide fire extinguishers in the B-17: one on the aft bulkhead of the navigator's compartment, one on the right rear bulkhead of the pilot's compartment, and the third on the forward bulkhead of the radio compartment.

To operate—Stand close to fire, raise horn and direct gas to the base of the fire holding onto the rubber insulated tubing. **Warning:** Do not grasp metal horn on top of cylinder; the white gas discharged is "dry ice" and will cause frostbite.

To shut off flow of gas, return horn to position at side of cylinder. Recharge cylinder after each use.

2. Two carbon tetrachloride fire extinguishers are also provided: one at the copilot's left under the seat and one aft of the main entrance door.

To operate—Turn handle and pump plunger, keeping stream full and steady. Stand as far as possible from the fire when using this extinguisher. Effective range is 20 to 30 feet. To shut off, push handle in and turn until sealing plunger is depressed. **Caution:** When sprayed on a fire carbon tetrachloride produces phosgene, an extremely poisonous gas, which can be harmful even in small amounts, and can prove fatal if inhaled. Do not stand near fire. Open windows and ventilators immediately after fire is extinguished.

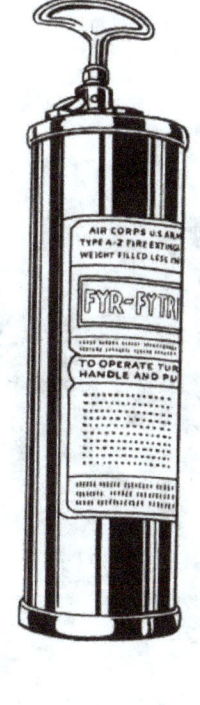
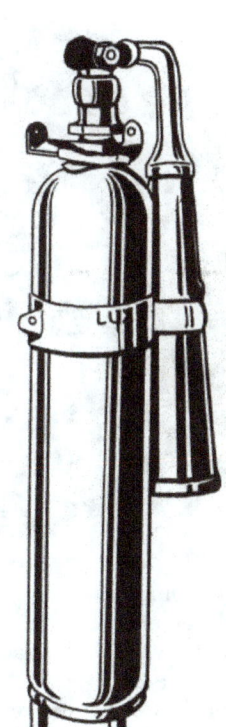

EMERGENCY SIGNAL EQUIPMENT

1. **Alarm Bells:** There are 3 alarm bells on the B-17 for use in emergencies. One is under the navigator's table, one above the radio operator's table, and the third in the tail compartment inside the dorsal fin. A toggle switch on the pilot's control panel controls them.

Operation—Stand by to abandon: Give three short rings. Abandon airplane: One long continuous ring.

2. **Phone Call:** A toggle switch on the pilot's control panel operates 4 amber phone-call signal lamps. Three of them are adjacent to the alarm bells and the fourth is in the tail gunner's compartment on the right side looking aft.

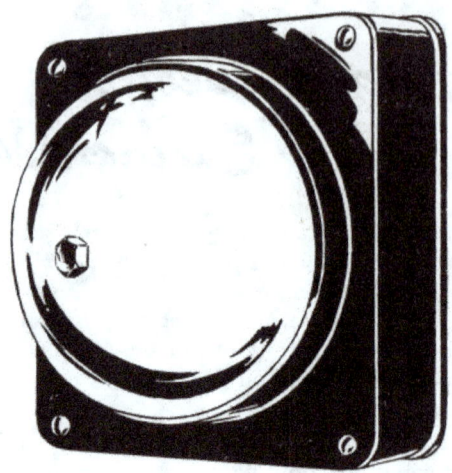

EMERGENCY RADIO TRANSMITTER

1. Some planes have a self-contained portable emergency radio transmitter, stowed on the forward bulkhead of the waist compartment. It is provided for operation anywhere away from the airplane. It is primarily for use in a life raft, but may be operated anywhere a kite may be flown or where a body of water may be found. It has a small parachute to permit dropping from the airplane from an altitude of 300 to 500 feet in an emergency.

2. When operated, the transmitter emits an MCW signal on the international distress frequency of 500 Kc. Automatic transmission of a predetermined signal is provided. Any searching party can make a homing on the signal with the aid of a radio compass.

3. No receiver is provided.

4. Complete operating instructions are contained with the equipment.

5. If emergency landing is made on water, the emergency radio set should be removed at the same time the life raft is removed. The set is waterproof and will float. Be sure the set does not float out of reach.

6. To bail out the emergency radio, tie the loose end of the parachute static line to any solid structure of the airplane and throw set through any convenient opening. **Be sure static line does not foul.**

FIRST-AID KITS

1. First-aid kits are on the bombsight storage box, in the navigator's compartment, on the wiring diagram box, on the back of the copilot's seat, and on the bulkhead forward of the lower turret.

2. If first-aid kits are not installed, it is necessary to obtain sufficient number before flight.

3. The first-aid kit, aeronautic, contains the following: tourniquet (1); morphine syrette (2); wound dressing, small (3); scissors (1 pair); sulfanilamide crystals, envelope (1); sulfadiazine tablets (1 box of 12 tables); burn ointment (1 tube) (boric or 5% sulfadiazine); eye dressing set; halazone tablets; 1-inch ad-

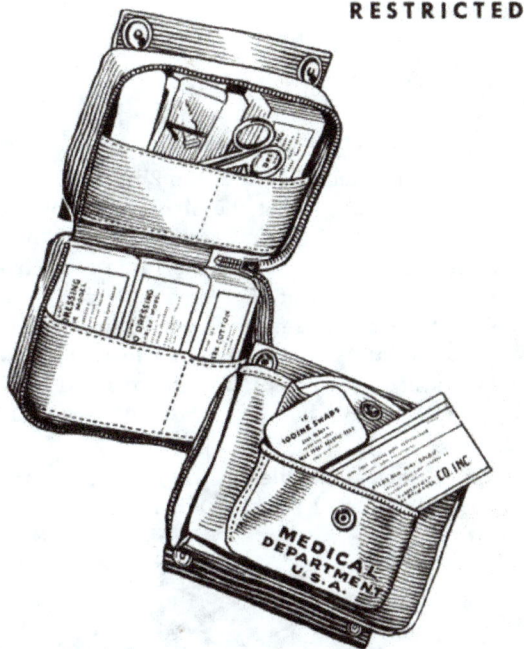

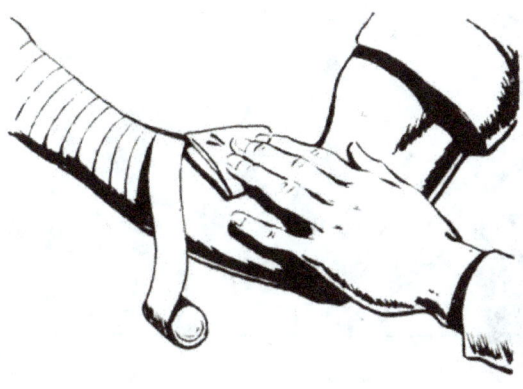

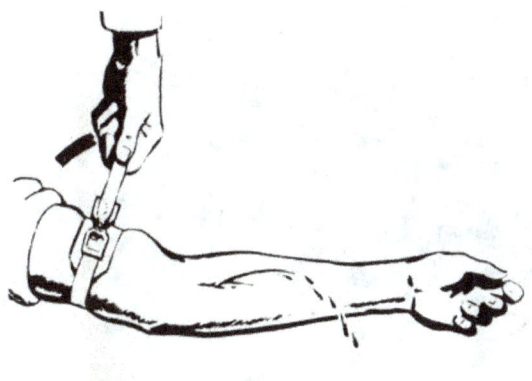

hesive compresses (1 box) (contents of small outer pocket); Iodine swabs (10) (contents of small outer pocket):

4. Use—In the case of a wound, first stop the flow of blood. The clothing should be cut away and a compress of wound dressing applied after the sulfanilamide powder has been sprinkled into the wound. If a firmly applied dressing will not stop the bleeding, or if there is actual spurting of blood from an artery, the tourniquet should be applied. A tourniquet must be released every 20 minutes and removed as soon as hemorrhage stops.

5. To relieve severe pain, open the small cardboard container and follow directions given there for the use of the hypodermic syrette of morphine. Do not hesitate to use the hypodermic to relieve suffering.

6. In case of head injury have the man lie quietly with head slightly elevated.

7. In the event of marked blood loss with shock and/or unconsciousness, have the man lie horizontally or lie with the head down, if possible.

8. An adequate supply of oxygen is doubly important in case of serious injury. Use it generously.

LIFE RAFT

An automatically ejected life raft (Type A-Z or A-3) is carried in each of the 2 life raft compartments in the top of the airplane aft of the top turret. The 2 life rafts are released by 2 pull handles, near the ceiling of the radio compartment just forward of the removable top window. These 2 release handles are clipped into a rack and safety-wired into place to avoid their being pulled by accident. To release a raft completely, pull the handle, hard, out about 12 inches.

The 2 release handles in the radio compartment are attached to the latch mechanism by cables. The functions of the latches are to keep the life raft compartment doors from opening at the wrong time and to insure operation in emergency release. A cable also connects the latch mechanism and the CO_2 bottle valve in the rafts.

Operation—A hard pull of about 12 inches on the release handles in the radio compartment causes the latch mechanism to release the raft compartment doors, and at the same time discharge CO_2 into the raft. Inflation of the raft forces it from the compartment into the water. A mooring line with a low breaking tension is provided to hold the raft in the vicinity of the aircraft. Accessories are provided for use while in the water.

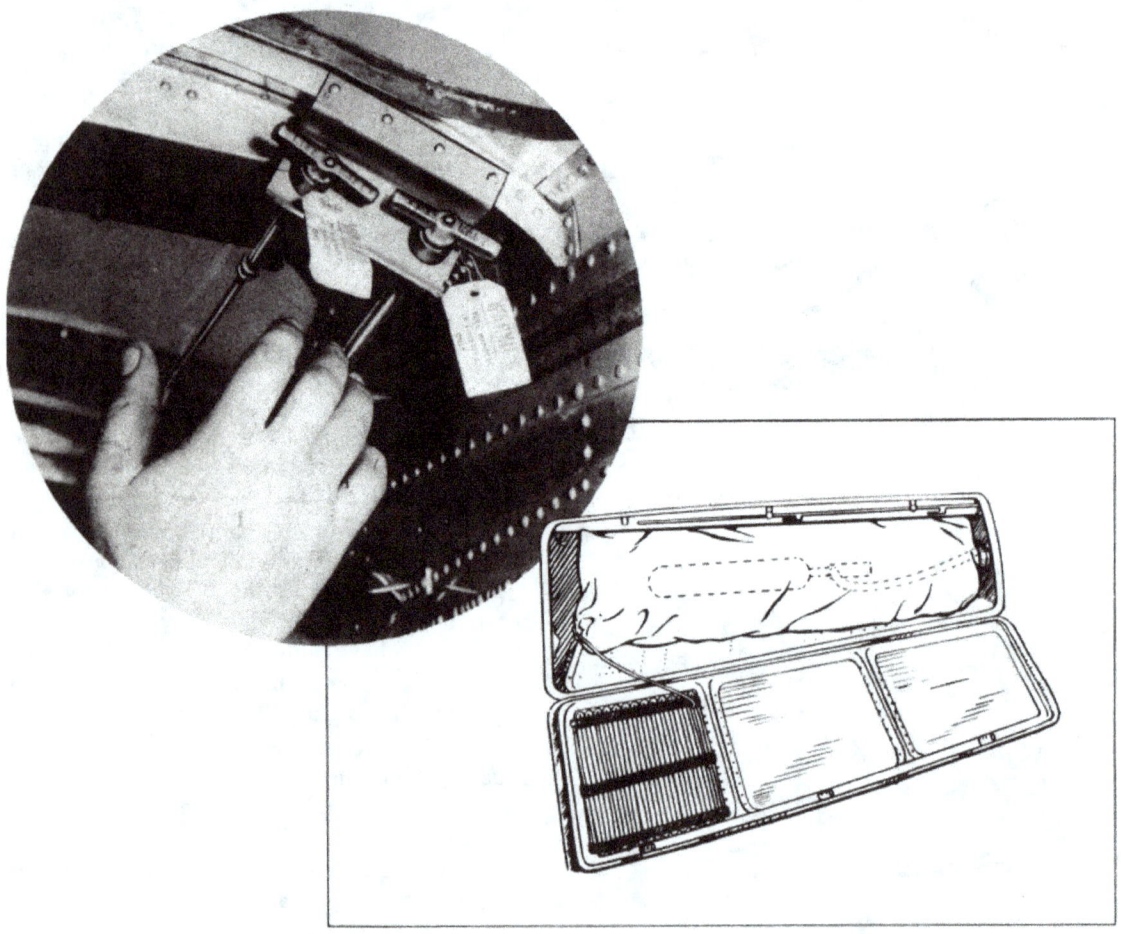

PYROTECHNIC PISTOLS

There is a pyrotechnic pistol in the cockpit behind the pilot's seat. Flares are generally mounted on the roof behind the pilot.

When radio communication is inadvisable or when radio equipment fails, brief coded messages may be sent with pyrotechnic signals. Do not use pyrotechnic signals to control important operations unless no other means is available. The various colored signals which are available for use with M2 and AN-M8 pyrotechnic pistols are assigned different meanings under a code that will be changed at frequent intervals in each edition of **Signal Operation Instructions**. The M11 red star parachute signal, however, is always used as a distress signal to be fired from the ground or from a life raft.

M2 Pistol

The M2 pyrotechnic pistol has a strong recoil. Use both hands to fire it if practicable. The signals themselves burn with an extremely hot flame; observe every reasonable precaution while handling or firing them.

1. Fire signals only from airplane in flight, with the exception of the M11 distress signal.
2. Point the pistol in such a way as to keep signals from striking any part of the plane.
3. If a signal fails to ignite on the first attempt, try at least twice more. If third or final try fails, keep the pistol pointed overboard and clear of all parts of the airplane for at least 30 seconds; then discard signal.
4. Discard a misfired signal, if possible, without handling the signal itself. One method is to hold the pistol over an opening in the airplane and release the cartridge by pressing on the latch and allowing the signal to fall clear under the force of gravity. The force of the air blast prevents holding the pistol outside most airplanes. **Be careful to prevent discarded signal from striking any part of the airplane.**
5. Do not discard misfired signals when flying over populated areas.
6. Fire the M11 distress signal as nearly straight up as is practicable.

AN-M8 Pistol

The AN-M8 pyrotechnic pistol is replacing the M2 pistol. It is fired by inserting and locking the barrel in a type M-1 mount. This mount is really a little door, fastened rigidly to the airplane, which permits the pistol barrel to extend through the airplane's outer skin. The mount absorbs the recoil of the pistol. Observe these precautions in using this pistol:

1. Place cartridge in chamber after pistol is inserted in mount, and only when immediate use is anticipated.
2. Since the pistol is cocked at all times when the breech is closed, **never leave a live signal in the pistol when it is removed from the mount.**

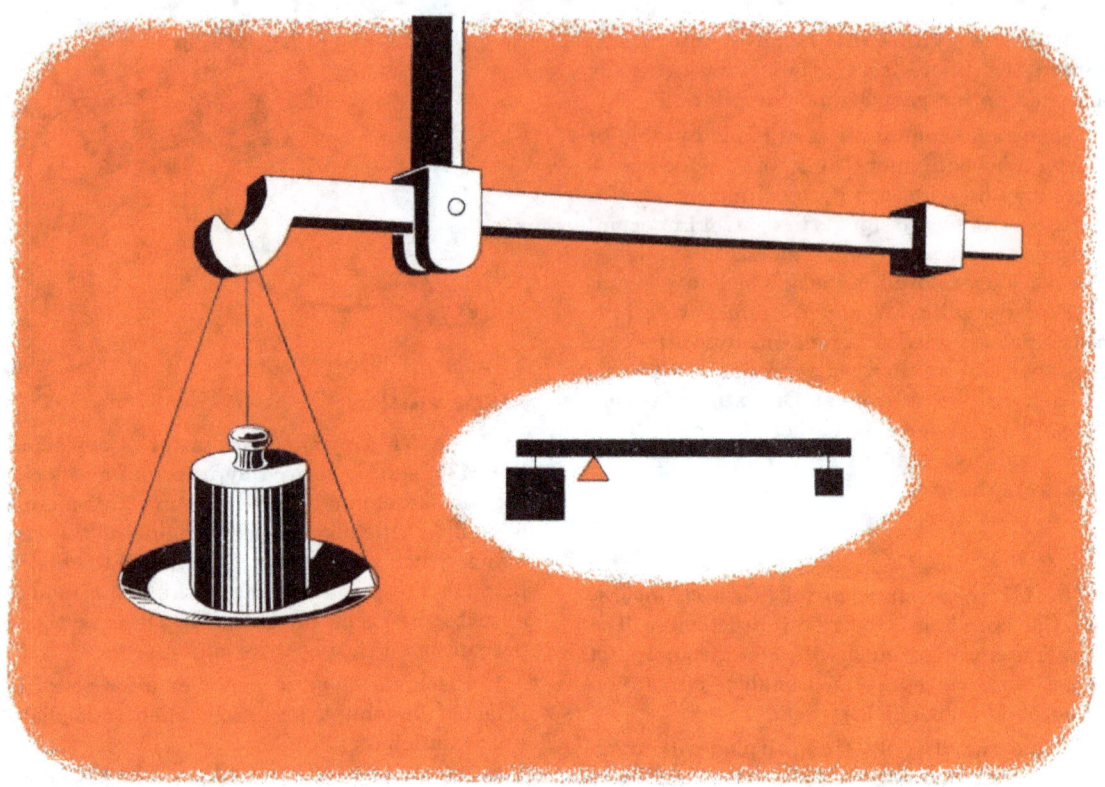

WEIGHT *and Balance*

The day when a pilot flew by the seat of his pants is past. One by one the decisions that were made by intuition or hunches have been taken over by an orderly system based on knowledge and understanding. The invariable result has been greater safety and operating efficiency.

The loading of aircraft, especially heavy aircraft, is no exception. The ever-changing conditions of modern airplane operation, resulting in more and more complex combinations of cargo, fuel, crew, and armament, have outmoded rule-of-thumb methods. The necessity for getting the utmost in efficiency out of any given flight has high-lighted the need for a precise system of control over the weight and balance of aircraft.

Improper loading, at best, cuts down the efficiency of an airplane from the standpoint of ceiling, maneuverability, rate of climb, and speed. At worst, it can be the cause of failure to complete a flight—or for that matter, failure even to start it—with probable loss of life and destruction of valuable equipment, because of abnormal stresses upon the airplane or because of changed flying characteristics.

EFFECTS OF IMPROPER LOADING

OVERLOADING

1. Causes a higher stalling speed.
2. Always results in lowering of airplane structural safety factors which may be critical during rough air or takeoffs from poor fields.
3. Reduces maneuverability.
4. Increases takeoff run.
5. Lowers angle and rate of climb.
6. Decreases ceiling.
7. Increases fuel consumption for given speed, which decreases the miles per gallon.
8. Lowers tire safety factors.

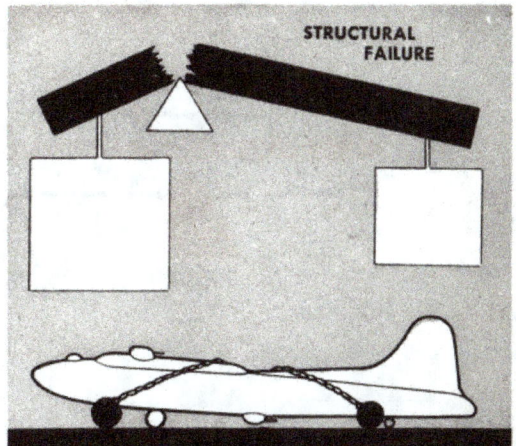

CG TOO FAR FORWARD

1. Increases fuel consumption (less range).
2. Increases power for given speed.
3. Tends to increase dive beyond control.
4. Might cause critical condition during flap operation.
5. Increases difficulty in getting tail down during landing.
6. Results in dangerous condition if tail structure is damaged or surface is shot away.

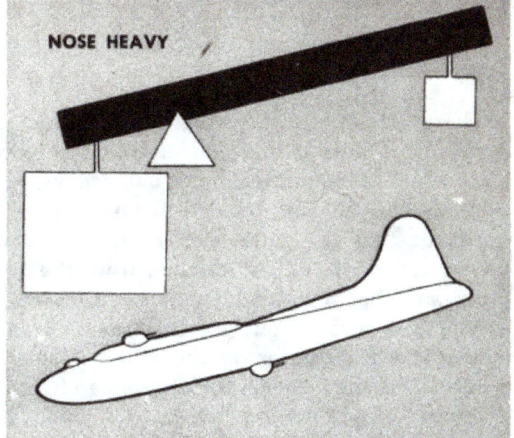

CG TOO FAR AFT

1. Creates unstable condition.
2. Increases stall tendency.
3. Definitely limits low power; might affect long-range optimum speed adversely.
4. Decreases speed.
5. Decreases range.
6. Increases pilot strain in instrument flying.
7. Results in a dangerous condition if tail structure is damaged or surface is shot away.

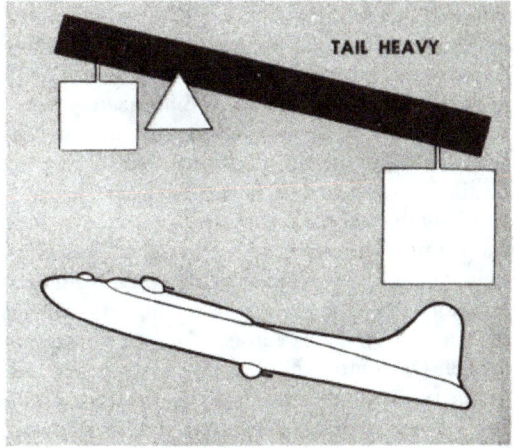

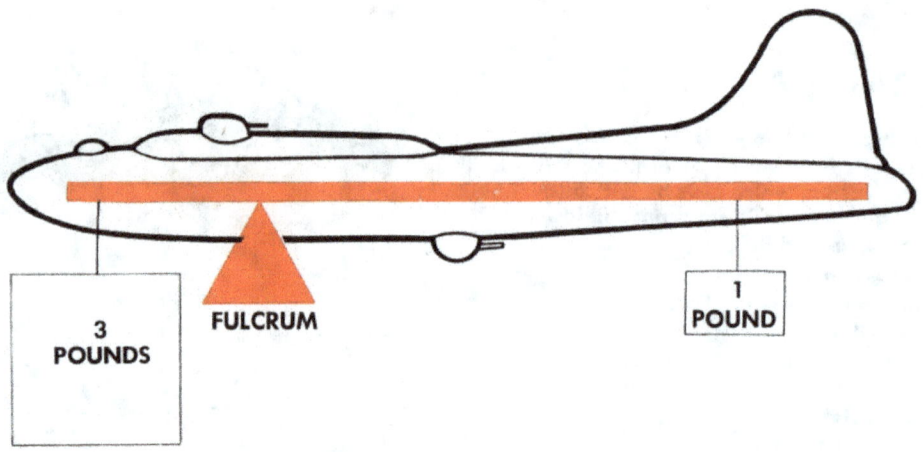

PRINCIPLES OF BALANCE

The theory of aircraft weight and balance is simple. It is that of the old familiar steelyard scale which is in equilibrium or balance when it rests on the fulcrum in a level position. It is apparent that the influence of weight is directly dependent on its distance from the fulcrum and that the weight must be distributed so that the turning effect is the same on one side of the fulcrum as on the other. A heavy weight near the fulcrum has the same effect as a lighter weight farther out on the bar. The distance of any object from the fulcrum is called its **arm**. This distance, or arm, multiplied by the weight of the object is its turning effect, or **moment**, exerted about the fulcrum.

Similarly, an airplane is balanced when it remains level if suspended at a certain definite point or ideal center of gravity (CG) location. Unlike a steelyard, it is not necessary that an airplane balance so that it is perfectly level, but it must be reasonably close to it. This allowable variation is called the CG range; the exact location, which is always near the forward part of the wing, is specified for each airplane model. Obtaining this balance is simply a matter of placing loads so that the average arm of the loaded airplane falls within this allowable CG range. Heavy loads near the wing location can be balanced by much lighter loads at the nose or tail of the airplane. The moments determine this exactly.

In practice, it has been found desirable to measure all distances from an arbitrary reference datum line at or near the nose of the airplane. By measuring arms in the same direction all moments become positive, thus eliminating possible errors in adding plus and minus moments that result from a reference datum line located within the limits of the airplane.

When the total moment about this reference datum line is divided by the total weight, the resulting arm is the distance to the center of balance, or center of gravity, from the reference datum line. This would be the location of the fulcrum as illustrated on the balanced steelyard scale. If the CG falls within the CG limits, expressed as forward limit and aft limit, the loading is satisfactory. If not, the load must be shifted until the CG does fall within the limits.

For flight, since the wing supports the airplane's weight, it is obvious that the CG must remain within safe allowable limits; otherwise, the tail surfaces could not properly control the path of flight. Limits are usually expressed as a percentage of the mean aerodynamic chord of the wing (% MAC). However, for weight and balance purposes, and in this manual, the limits are given in inches from the reference datum line.

To obtain the gross weight and the CG location of the loaded airplane, it is necessary first to know the basic weight and the CG location of the airplane. This may be found by weighing the airplane. This weighing should be done with the airplane in its basic condition; that is, with fixed normal equipment which is actually present in the airplane, less fuel.

When the weight, arm, and moment of the basic airplane are known, it is not difficult to compute the effect of fuel, crew, cargo, armament, and expendable weight as they are added. This is done by adding all the moments of these additional items to the total moment found by weighing the airplane and dividing by the sum of the basic weight and the weight of these additional items. This gives the CG for the loaded airplane. This calculation can be performed by arithmetic, with loading graphs, or with a balance computer.

LOADING GRAPHS

Loading graphs and detailed instructions for their use are included in Section 7 of AN 01B-40 **(Weight and Balance Data)** a copy of which must be kept in the data case of the airplane at all times. These loading graphs provide an easy means of determining the loaded CG position of the airplane. They are intended for use when the balance computer is not available.

BALANCE COMPUTERS

To simplify the work of determining the loaded CG of the airplane, a balance computer is provided for each B-17 airplane and certain other types of aircraft, such as transports and patrol bombers, which may be easily unbalanced by improper loading and which carry such a large number of variable load items that calculation of their loaded CG by arithmetic or with the aid of loading graphs might be a somewhat lengthy and tedious process. There have been several types of computers used for this purpose. However, the load adjuster has been adopted as the standard computer for both the Army and the Navy. Instructions for using the load adjuster are included in Appendix I of AN 01B-40.

Definitions

The following definitions will serve as standardized terminology for all data in the practical application of this system. It is important to know them thoroughly.

Weight—The weight is 16 ounces per pound, avoirdupois weight. All weights are to be calculated to the nearest whole pound.

Basic Weight—The weight of the airplane, including all equipment that has a fixed location and is actually present in the airplane; that is, air frame; power plants and accessories; trapped fuel and oil; full hydraulic, cooling and anti-icing fluid systems and reservoirs; armor plate, ordnance (less ammunition and bombs); chemical, navigation, oxygen, pyrotechnics, and radio equipment. It never included items commonly referred to as disposable.

Note: The basic weight of an airplane varies with modifications and changes in the fixed equipment. This is not to be confused with empty weight, which is a dry weight with certain contract equipment only. The term basic weight, when qualified with a word indicating the type of mission, such as "basic weight for combat, for ferry, for transport, etc.," may be used with directives stating what the equipment shall be for these missions; for example, extra fuel tanks and various items of equipment installed for long-range ferry flights but not normally carried on combat missions which will be in "Basic Weight for Ferry" but not in "Basic Weight for Combat."

Gross Weight—The total weight of an airplane and its contents.

Reference Datum Line—An imaginary vertical line at or near the nose of the airplane. Its location is chosen by the manufacturer as a standard line from which all horizontal distances are measured for balance purposes. Diagrams of each airplane show this reference line as zero.

Arm—For balance purposes, arm is the horizontal distance in inches from the reference datum line to the CG of the item.

Moment—The weight of an item multiplied by its arm.

Average Arm—Average arm or location is obtained by adding the weights and the moments of a number of items and dividing the total moment by the total weight.

Basic Moment—The sum of the moments of all items making up the basic weight. When using data from an actual weighing of an airplane, the basic moment is the sum of the moments around the reference datum line. For simplicity, it is permissible to divide the moment by a constant so as to reduce the number of digits. If this is done, the same constant must be used consistently for all computations, and must be indicated in the moment column on charts A, B, and C in Form F.

Center of Gravity—The point about which an airplane would balance if suspended. Its distance from the reference datum line is found by dividing the total moments by the gross weight of the airplane.

CG Limits—The range of movement which the CG can have without making the airplane unsafe to fly. It is determined by actual test flights. The CG of the loaded airplane must be within these limits at takeoff, in the air and on landing. In some special cases a landing limit is specified. On loading graphs the CG limits are indicated by CG limit lines. In all cases, the CG condition should be checked for landing without fuel and bombs.

Loading Range—The safe CG location under any load condition. It is shown on the balance computer as the white section labeled "Loading Range."

Tare—Weight of equipment necessary for weighing the airplane (chocks, blocks, slings, jacks, etc.) which is included in the scale readings but is not part of the basic weight.

Balance Computer Index—A number representing the amount which, when considered in conjunction with the weight, gives the CG position.

Index

Accidental Unfeathering 142
After-Landing Check 98
Airplane Commander 13-25
Alternating Current 165
Approach, Final 94-95
    Power 95
    Power-Off 94-95
Armor 42-43
Automatic Pilot ....(See Autopilot)
Autopilot, the C-1....................183-190
    Emergency Use of....................187
    for Bomb Approach.................20-22
    Ground Checklist 190
    How to Operate..................185-188
Auxiliary Power, Use of..................165

B-17, History 5-12
    in Combat 9-12
B-17G, General Description...........43-45
Bail Out, How to....................148-151
Balance Computer 201
    Principles of 200-201
Ball Turret, Dropping the..........134-135
    General Description 40
    in Emergency Landings...........136-137
Banks, Load Factor in.................... 88
Before-Landing Check 93-94
Before Takeoff 69
Boeing 299 5
Bomb Bay, General Description........... 39
Bomb Bay Doors, Emergency Operation..133
Bombardier, Duties of.................18-22
Bombardier-Pilot Relationship 18-22
Booster Pump 61, 159
Brakes 164
Brake Operation With Hydraulic Pump
    Failure 139

C-1 Autopilot................(See Autopilot)
Carburetor Ice, Prevention in Flight...105-106
Celestial Navigation 17
Center of Gravity in Heavy Loads......... 91
    in Improper Loading.................199
    Limits, Determination of.........200-202
Changes in Equipment, B-17F and G....... 45
Characteristics, Flight 88-91

Checklist, Approved B-17F............55-56
Chin Turret, in B-17G.................43-44
Circular Error 22
Climb, Angle of......................... 74
    Auto-Rich Mixture for................ 74
    Decreasing Air Temperature in.....74-75
    Decreasing Atmospheric Pressure in.... 75
    Use of Turbo-supercharger in......... 75
Climbing 72-87
    Effects of Altitude in................. 74
    Engine Heat in....................... 74
    Power Required and
    Power Available (Chart).............. 73
    Power Settings for................... 72
    on Instruments 72
Cockpit Checklist 54-56
Cold Weather Operation............102-108
    Warm-up 103
Commander, Airplane 13-25
Communication 176-182
Compass, Gyro Flux Gate...........191-192
Control Panel and Pedestal...........34-35
Cowl Flaps, Effects of................... 88
Crash Landings 156
Crew Discipline 13-14
    Training 14
Crosswind Landing 95-96
    Takeoff 71
Cruising 81-84
    Long-Range 84
    Maximum Endurance 84

Dead Reckoning 16
Decreasing Air Temperature in Climb...74-75
Decreasing Atmosperic Pressure in Climb.. 75
De-icer System 107-108
Detonation and Pre-ignition............85-87
Dimensions 27
Discipline, Crew 13-14
Ditching 152-155
Ditching, Crew Duties in............153-154
    Crew Positions for................154-155
    Wind Speed in......................155
Dives 91
Dropping the Ball Turret in Flight.....134-135

Electrical System 165-168
Electronic Turbo-supercharger Control.169-171
Emergency Brake System................164
    "   Equipment 193-197

" Feathering 141-142
" Hydraulic System 139
" Operation of Bomb Bay Doors........133
" Operation of Landing Gear..........133
" Operation of Wing Flaps............133
   Radio Transmitter 194
   Signal Equipment 194
   Unfeathering 143
End of Mission........................ 98
Engine Failure:
   One-engine failure on takeoff..........144
   Two-engine failure on takeoff..........145
   Go-around with one engine out.........146
   Two-engine landing 146-147
   Single engine operation................147
Engine Heat in Climb................... 74
   Primer 159
   Run-up 68
   Section Fire Extinguisher..............193
Engineer, Duties of..................... 24
Engines, General Description............ 30
Equalizer Coils 166
Evasive Action 21

Feathering 140-144
   Practice 144
   Procedure 140-141
   System, Failure of................142-143
Final Approach 94-95
Fire, Engine, in Flight..................129
   Engine, on Ground...................129
   Extinguisher, Engine Section..........193
   Extinguishers 193
Fires in Flight.....................128-129
First-Aid Kits 195
Flight Characteristics 88-91
   Performance Record (Form)..........82-83
Formation Flying 119-127
   Flying, Tips on....................127
   Landing Procedure 125-127
   Takeoffs 120-121
Frequency Meter 181
Fuel Pump, Engine-Driven..............159
   Capacity 157
   Shut-off Valves 159
   System 157-161
   Transfer Pump 160-161
   Transfer Selector Valve................160
Fuselage, General Description.......... 28

General Description of B-17F..........26-43
Generator Systems, Checking and Adjusting 166
Generators 165
Go-Around 97
   With One Engine Out................146
Grade 91 Fuel, Note on Use of..........85-87
   Power Settings for..................85-87
Grade 100 Fuel, Power Settings for......85-87
Gravity, Center of (See Center of Gravity)
Gunners, Duties of.................... 25
Gyro Flux Gate Compass.............191-192

Hand Fuel Transfer Pump................161
Heating and Ventilating System.......174-175
Heavy Loads 91
Hot Weather Tips.....................109
Hydraulic Pump Failure, Brake Operation
   With 139
   System 163-164
   System, Emergency 139

Ice, Emergency Removal................106
Icing, on Aircraft..................106-108
Inspections and Checks..............46-56
Instrument Calibration 17
   Check 64
   Panel 38B
Intercoolers, Function of................169
   in Climbing 75
   in Landing 93
   in Starting 57
Interior of Airplane, General Description.33-43
Inverter Check 61

Landing 92-98
" Disabled Aircraft 136-137
" Gear, Emergency Operation of........133
" Gear, Main 31
" Roll 97
" Crash 156
" Crosswind 95-96
" in Strong Wind................... 95
" Maximum Performance 131-132
" Night 101
" No-flap 132
" on One Flat Tire..................136
" Two-Engine 146-147
" With Bent Drag Link................136
" With Broken Drag Link..........136-137
" With Cracked or Wobbling Wheel.....136

Entry	Page
Leveling Off	78
Liaison Transmitter	178
Life Raft	196
Load Factor in Turns and Banks	88
Loading Graphs	201
Effects of Improper Loading	199
Long-Range Cruising	84
Main Landing Gear	31
Marker Beacon	181
Maximum Endurance Cruising	84
Performance Landing	131-132
Performance Takeoff	130-131
Mean Aerodynamic Cord	91, 201
Mission, End of	98
Navigation	15-18
Navigator, Duties	15-18
Navigator-Pilot Relationship	15-18
Night Flying	99-101
Flying, Illusions in	99-100
Landings	101
Takeoff	101
Taxiing Precautions	101
Vision, Precautions	100
Vision, Tips on	100
No-flap Landing	132
Nose Section, General Description	33
Note on Use of Grade 91 Fuel	85-87
Oil Cooler	161
Dilution	103-104
System	161-162
One-Engine Failure on Takeoff	144
Overpriming	103
Overspeeding Turbos	138
Oxygen	110-118
PDI, Flying the, Manually	189
for Bomb Approach	20-22
Pilot-Bombardier Relationship	18-22
Pilotage	15
Pilot's Compartment	33-38B
Control Panel	37
Pilot-Navigator in Flight	17
Post-flight Critique	18
Preflight Planning	17
Relationship	15-18
Pilot's Operational Equipment	34-38B
Power Approach	95
Changes, Sequence of	76-77
Increase, Sequence for	76
Plant, General Description	30
Reduction, Sequence for	77
Settings for Grade 100 and 91 Fuel	85-87
Power-off Approach	94-95
Practice Feathering	144
Preheating	102
Pre-ignition, Detonation and	85-87
Primer, Engine	159
Propeller, Anti-icer System	107
Propellers, General Description	30
Runaway	137-138
Synchronization of	80
Pyrotechnic Pistols	197
Radio Compartment, General Description	40
Compass	178-179
Equipment	176-182
Equipment, Emergency Operation of	182
Navigation	17
Operator, Duties of	23
Set, SCR522A (VHF)	180-181
Transmitter, Emergency	194
Recovery From a Stall	90
Reverse Current Relay	166
Rough Air Operation	89
Runaway Propellers	137-138
Running Takeoff	71
Run-up Procedure	68
Signal Equipment, Emergency	194
Single Engine Operation	147
Spins	91
Stall Recovery	90
Stalls	89
Starting Procedure	57-64
Strong Winds, Landing in	95
Supercharger Regulator Operation	169
Synchronizing Propellers	80
Tail Assembly, General Description	29
Tail Gunner's Compartment	41
Tailwheel, General Description	32
Takeoff, Crosswind	71
Maximum Performance	130-131
Night	101
Running	71
Technique	67-70
Taxiing Technique	65-66
Throttle Technique	69
Traffic Pattern	92

Training, Crew 14
Trim Tabs, Effects of..................... 88
Trimming 79
Turns, Flight Characteristics in........... 88
    Load Factors in...................... 88
Two-Engine Failure on Takeoff...........145
    Landing 146
Turbos, Overspeeding 138
Turbo-supercharger Control, Electronic 169-171
    Use in Climb....................... 75-76
Turbo-superchargers 169-173
    Use of 172-173
Turbo Surge 173
    Waste Gate, Closed....................173

Unfeather, How to......................143
Unfeathering, Accidental 142

Voltage Regulator, Function of...........165

Waist Section, General Description........ 41
Weight 27
    and Balance 198-202
Windshield Anti-icer System.............107
Wing Flaps, Emergency Operation of.....133

X-B17 5

Y1B-17 7
Y1B-17A 7

www.ingramcontent.com/pod-product-compliance
Lightning Source LLC
Chambersburg PA
CBHW080024130526
44591CB00037B/2666